John H. Lowenstein

Pseudochaotic Kicked Oscillators

Renormalization, Symbolic Dynamics, and Transport

John H. Lowenstein

Pseudochaotic Kicked Oscillators

Renormalization, Symbolic Dynamics, and Transport

With 86 figures, 9 of them in color

高等教育出版社
HIGHER EDUCATION PRESS

Springer

Author
John H. Lowenstein
Department of Physics
New York University
2 Washington Place
New York, NY 10003, U.S.A.
Email: john.lowenstein@nyu.edu

ISBN 978-7-04-032279-8
Higher Education Press, Beijing

ISBN 978-3-642-28153-2 ISBN 978-3-642-28154-9 (eBook)
Springer Heidelberg New York Dordrecht London

Library of Congress Control Number: 2012930642

Printed on acid-free paper

Springer is part of Springer Science+Business Media (www.springer.com)

Dedicated to the memory of George Zaslavsky

Preface

The physics of a pseudochaotic kicked oscillator could hardly be simpler: a one-dimensional harmonic oscillator, subjected to impulsive kicks in resonance with the natural frequency, with the amplitude of the kicks a periodic, piecewise linear function of the position. As with other nonlinear systems, the simple dynamics can produce, over a long time, exceedingly complex behavior, typically too complex for meaningful mathematical analysis or even reliable numerical experiments. Fortunately, a special choice of parameters provides the key to a remarkably detailed understanding of the long-time asymptotics. That key is *renormalizability*. To show how renormalizability leads to a wealth of exact results, as well as powerful computational tools for exploring long-time asymptotics, has been my principal objective in writing this monograph. The exposition is reasonably self-contained, with references to the literature where supplementary details are needed. The methods used are largely traditional ones which should be accessible to readers with a modest familiarity with the dynamics of low-dimensional Hamiltonian systems. No previous acquaintance with pseudochaos is assumed.

This book weaves together a number of important threads drawn from a long, productive, and continuing collaboration with Franco Vivaldi at Queen Mary, University of London. Our research program, initiated in 1995, has resulted in a number of published articles on the topic of piecewise isometries and their various manifestations, including Hamiltonian round-off, interval exchange transformations, kicked-oscillator models, flights, and pseudochaos. Over the years, other researchers have contributed significantly to the enterprise, notably Gullaume Poggiaspalla, Konstantin Kouptsov, Sangtian Liu, and Spyros Hatjispyros.

In writing the book, I have drawn liberally from the published results, tying them together and also adding considerable new material, and new perspectives, in order to fashion a coherent whole. Central to the story is the unifying role of the symbolic dynamics. A detailed description of the latter, including the previously unpublished derivation of admissibility rules, appears in Chapter 3. In addition, the discussion of transport in Chapter 6 is entirely new. Finally, Chapter 7 integrates the results on Hamiltonian round-off, originally published during 1997—2000, into the full story, with the inclusion of a new numerical experiment to provide fresh evidence for some of the main conclusions.

Over the years, my collaborators and I have benefited from generous short-term support of our research from the Engineering and Physical Sciences Research Council (EPSRC) and The Royal Society. The continuing hospitality and support of the Department of Physics, New York University and the School of Mathematical Sciences of Queen Mary, University of London, have been crucial to both the research program and the writing of this book.

Some of the figures in the book have been copied, with permission of the publishers, from articles in the journals *Physics Reports* (Elsevier), *Nonlinearity* (IOP), and *Communications in Nonlinear Science and Numerical Simulation* (Elsevier).

Portions of the first five chapters were originally included in a series of pedagogical lectures at the National University of Singapore (NUS) in August, 2006. I very much appreciate the hospitality and financial support provided by the Institute of Mathematical Sciences of NUS at that time.

Finally, I want to thank Valentin Afraimovich and Albert Luo for their advice and encouragement throughout the writing of this book.

New York, August, 2011 *John H. Lowenstein*

Contents

Chapter 1
Introduction

In this initial chapter, we introduce a class of one-dimensional kicked oscillator models whose crystalline symmetry and renormalizability will allow us, in the course of this book, to explore the rich and varied relationships between dynamical self-similarity at the local level and transport on the infinite phase plane.

1.1 Kicked oscillators

The dynamics of a periodically kicked oscillator is governed by a Hamiltonian of the form (in convenient units)

$$H(x,y) = \frac{1}{2}(x^2+y^2) - F(x)\sum_n \delta(t-2\pi n\rho), \tag{1.1}$$

where the rotation number ρ is the number of instantaneous kicks per natural period. It is assumed that the kicks are in resonance with the unperturbed oscillations, so that ρ is a rational number, and that the derivative $f(x) = F'(x)$ is a periodic function of the oscillator position. Hamilton's equations of motion take the form

$$\dot{x} = \frac{\partial H}{\partial y} = y, \qquad \dot{y} = -\frac{\partial H}{\partial x} = -x + f(x)\sum_n \delta(t-2\pi\rho n). \tag{1.2}$$

Between successive kicks, the system undergoes free oscillation, depicted in the x,y phase space as uniform clockwise motion on a circular arc of arbitrary radius and angle $2\pi\rho$. This is followed by an instantaneous momentum shift $y \to y+\Delta y$, where Δy is given by the kick function $f(x)$. In Fig. 1.1, we illustrate such a phase-space orbit for a 4-fold resonance and a sinusoidal kick function. The example is typical of kicked-oscillator models, introduced in the 1980's by Zaslavskii et al. (1986, 1991) to model the interaction of electromagnetic waves with gyrating charged particles in a plasma, and more abstractly, to illustrate the dynamical generation of crystalline and quasicrystalline order in 2-dimensional phase space.

1.2 Poincaré sections

The simplicity of the motion between kicks in Fig. 1.1(a) suggests that it might be advantageous to adopt a stroboscopic point of view, regarding the essence of the dynamics to be a discrete (Poincaré) map W connecting the phase-space points (x_n, y_n) at which the kicks are initiated. Explicitly, we have

$$W : \mathbb{R}^2 \to \mathbb{R}^2, \qquad \begin{pmatrix} x \\ y \end{pmatrix} \mapsto \begin{pmatrix} \cos 2\pi\rho & \sin 2\pi\rho \\ -\sin 2\pi\rho & \cos 2\pi\rho \end{pmatrix} \begin{pmatrix} x \\ y + f(x) \end{pmatrix}. \tag{1.3}$$

As is typical of dynamical systems with one degree of freedom and periodic forcing, the stroboscopic phase space (Poincaré section) is partitioned into disjoint invariant subsets. These subsets may be collections of points (periodic orbits), curves (quasi-periodic orbits) and more complicated sets (stochastic layers) populated in part by chaotic orbits. The discreteness of the dynamical map makes it especially easy and efficient to visualize such sets via computer-assisted iteration.

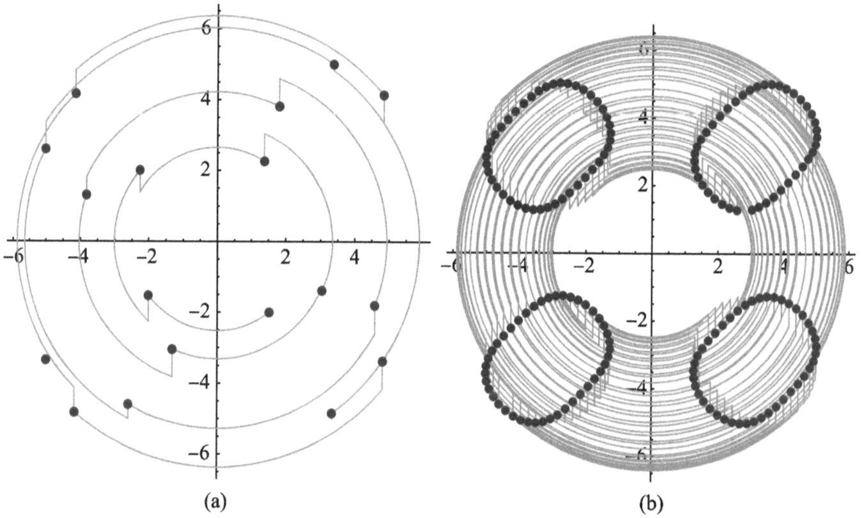

Fig. 1.1 Quasiperiodic orbits of the sinusoidal kicked oscillator with $F(x) = -a\cos x$, $a = 0.8$, calculated over **(a)** 5 and **(b)** 50 oscillation periods. Points of the Poincaré section are shown as large dots.

Figure 1.1(b) follows the orbit of Fig. 1.1(a) through 200 kick periods (50 oscillation periods). The phase-space orbit occupies an approximately annular region with numerous self-intersections. Without emphasizing (with dots) the points of the Poincaré section, the pattern would be confusing. Further iteration would make matters worse, since we would eventually be left with a featureless annular region overlapping those of nearby orbits. The Poincaré section, on the other hand, reveals the topological simplicity of the orbit, which, viewed stroboscopically, fills out quasiperiodically four symmetrically placed closed curves. Choosing other initial conditions reveals a "phase portrait" in which the curves of the example are part of an infinite family of orbits circulating around the periodic points (π,π), $(\pi,-\pi)$, $(-\pi,-\pi)$, $(-\pi,\pi)$, as shown for example, by Fig. 1.5(b) in Sect. 1.4.

For $\rho = 1/4$ in general, the Poincaré map (1.3) simplifies to

$$W\begin{pmatrix} x \\ y \end{pmatrix} = \begin{pmatrix} 0 & 1 \\ -1 & 0 \end{pmatrix}\begin{pmatrix} x \\ y+f(x) \end{pmatrix} = \begin{pmatrix} y+f(x) \\ -x \end{pmatrix} \tag{1.4}$$

For reasons which will become apparent below, the map W of (1.4) is often referred to the 4-*fold web map*, and we will use this terminology in the remainder of the book.

1.3 Crystalline symmetry

The 4-fold web map provides a simple and elegant theoretical laboratory for studying transport in a low-dimensional Hamiltonian system. This is due in large part to its *crystalline symmetry*. Specifically, suppose that the kick function $f(x)$ has period τ, and that , for $\tau > 0$, $f(x)$ is continuous from the right on $[0,\tau)$, while, for $\tau < 0$, $f(x)$ is continuous from the left on $(\tau,0]$. Then every point of the real plane can be uniquely decomposed as the sum of a *local* vector $\mathbf{u}$ in the *fundamental domain*

$$\Omega = \begin{cases} [0,\tau)^2 & \tau > 0, \\ (\tau,0]^2 & \tau < 0, \end{cases}$$

and a *global* vector $\tau\mathbf{m}$, $\mathbf{m} = (m,n) \in \mathbb{Z}^2$.

From (1.4), we have, for all $\mathbf{u} \in \Omega, \mathbf{m} \in \mathbb{Z}^2$,

$$W(\mathbf{u}) + \tau\mathbf{I}\cdot\mathbf{m}, \qquad \mathbf{I} = \begin{pmatrix} 0 & 1 \\ -1 & 0 \end{pmatrix}. \tag{1.5}$$

Since $\mathbf{I}^4$ is the identity, we get from (1.5) the discrete translation invariance of the fourth iterate of W: for all $\mathbf{x} \in \mathbb{R}^2$, $\mathbf{m} \in \mathbb{Z}^2$,

$$W^4(\mathbf{x} + \tau\mathbf{m}) = W^4(\mathbf{x}) + \tau\mathbf{m}. \tag{1.6}$$

Figure 1.3 illustrates the decomposition (1.5) for a hypothetical choice of $f(x)$ depicted in Fig. 1.2.

To take full advantage of the local-global decomposition, we now introduce the piecewise continuous *local map* $K : \Omega \to \Omega$, defining, for all $\mathbf{u} = (u,v) \in \Omega$,

$$K(\mathbf{u}) = W(\mathbf{u}) - \tau\mathbf{d}(\mathbf{u}), \qquad \mathbf{d}(\mathbf{u}) = (\lfloor \tau^{-1}(v + f(u)) \rfloor, -1) \in \mathbb{Z}^2,$$

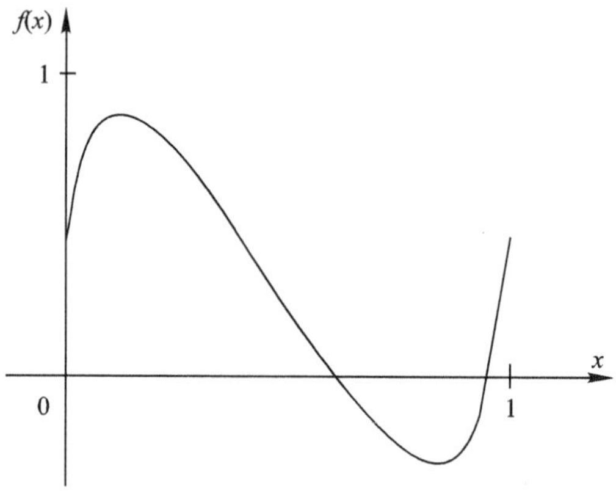

Fig. 1.2 Graph of some kick function $f(x)$ on $[0,1)$.

so that, for all $\mathbf{u} \in \Omega$, $\mathbf{m} \in \mathbb{Z}^2$,

$$W(\mathbf{u} + \tau \mathbf{m}) = K(\mathbf{u}) + \tau \tilde{L}_{\mathbf{u}}(\mathbf{m}),$$

with the lattice isometry (rotation composed with lattice translation)

$$\tilde{L}_{\mathbf{u}}(\mathbf{m}) = \mathbf{I} \cdot \mathbf{m} + \mathbf{d}(\mathbf{u}).$$

The action of K on Ω, for the example of Fig. 1.3, is shown in Fig. 1.4. We note that for piecewise continuous $f(x)$, the unit square Ω is partitioned into regions Ω_i, $i = 1, 2, \ldots, \nu$ on which the lattice translations $\mathbf{d}(\mathbf{u})$ are constant. Thus

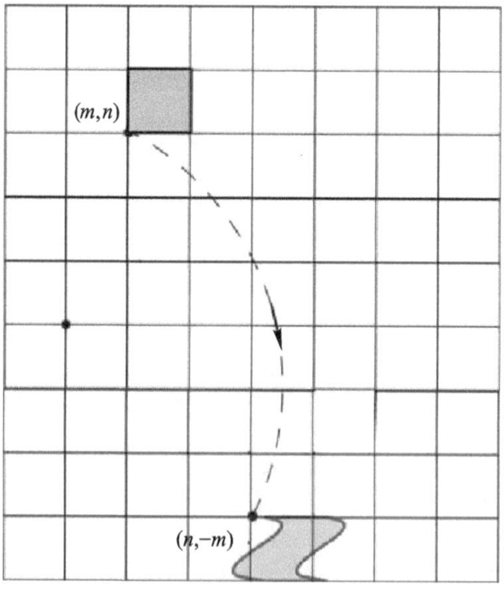

Fig. 1.3 Action of $W(\mathbf{x})$.

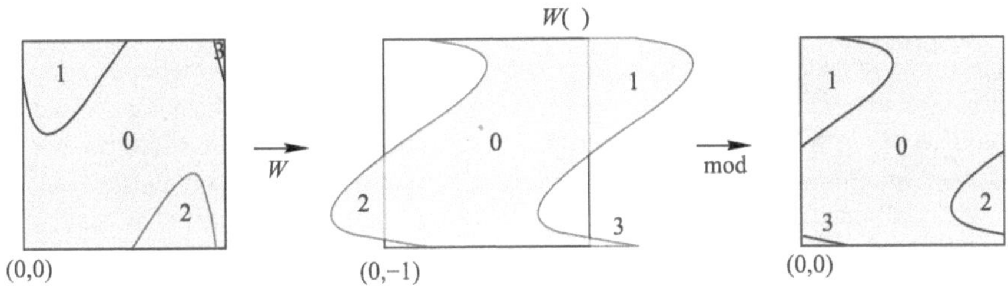

Fig. 1.4 Construction of $K(\mathbf{u})$ for the map of Fig. 1.3.

$\tilde{L}_{\mathbf{u}}$ depends on $\mathbf{u}$ only through the index $i(\mathbf{u})$, $u \in \Omega_{i(\mathbf{u})}$, and we can express the local-global decomposition of $W(\mathbf{x})$ as

$$W(\mathbf{u}+\tau\mathbf{m}) = K(\mathbf{u}) + \tau L_{i(\mathbf{u})}(\mathbf{m}), \tag{1.7}$$

where

$$L_{i(\mathbf{u})} = \tilde{L}_{\mathbf{u}}.$$

How does the crystalline symmetry of the dynamics, expressed through (1.7), facilitate the investigation of chaotic transport? For the latter, the asymptotic long-time behavior mimics the effects of a random walk on the plane, in which successive steps are dictated by the results of independent coin tosses. Clearly it is the deterministic chaos of the local map K which plays the role of the coin tosses, producing a "code" sequence $i_1, i_2, \ldots$, which in turn determines a sequence of steps on the infinite lattice. For asymptotically long times, it is the statistical distribution of the code-driven lattice coordinates which may in some sense exhibit diffusive behavior.

1.4 Stochastic webs

Although the main focus of this book is on maps which possess only some of the features of true chaos (hence the term pseudochaos), it is important, to properly understand the motivation for this work, that we focus first on the chaotic *stochastic web map* with kick amplitude

$$f(x) = a\sin x.$$

While the quasiperiodic orbits shown in Fig. 1.1(b) are restricted to only four cells, the same is not necessarily the case for chaotic orbits originating in the vicinity of one of the points $(m\pi, n\pi)$, $m+n$ odd. The linearized map approximating W^4 there is a 2×2 matrix, with one real eigenvalue greater than 1, the other less than one, so that these points are saddle points. By selecting initial points near one of these saddle points, say $(\pi, 0)$, it is easy to simulate numerically the orbits of the stochastic layer in which it is embedded. Each saddle point acts as a kind of random gate: on approaching it, an orbit must "decide" whether to turn right or left. The "decision" is deterministic, of course, but extremely sensitive to initial conditions whenever the orbit passes very close to a saddle point. The effect is very much like a punctuated random walk, with the orbit point moving stochastically from one saddle point's vicinity to another's after circulating locally. For ensembles of extremely long orbits, the local sensitivity to initial conditions translates into asymptotic behavior, in a statistical sense, resembling diffusion. For relatively small parameter a, the wandering chaotic orbits are restricted to a narrow web-like region connecting the saddle points at $(m\pi, n\pi)$, $m+n$ odd (see Fig. 1.5).

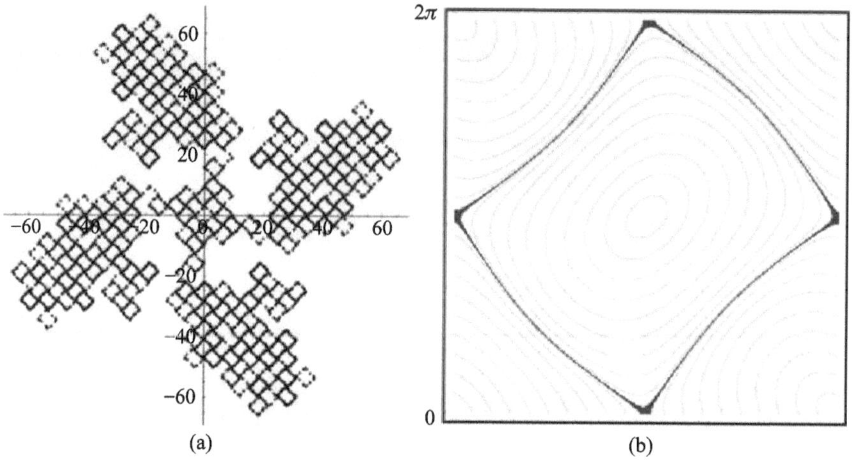

Fig. 1.5 (a) Chaotic orbit of the $a = 0.8$ sinusoidal kicked-oscillator map W (10^4 iterations). **(b)** Same orbit as (a), folded into the fundamental cell, showing structure of the stochastic web. Also shown are selected quasiperiodic orbits surrounding the fixed points at the center and at the corners of the cell.

Stochastic webs are present, not just for $\rho = 1/4$, but quite generally for sinusoidal kick function, rational ρ, and sufficiently small a (Zaslavskii et al., 1991). However, the cases $\rho = p/q$, $q = 3,4,6$ are the only cases where the 2-dimensional phase space has a periodic tiling by a finite number of cell shapes. This is a well-known constraint of 2-dimensional crystallography. On the other hand, there do exist mathematical patterns which exhibit quasi-crystalline structure (the Fourier transform of the density of points is point like and invariant under q-fold or $2q$-fold rotations). It is an empirical fact, based on numerical simulations, that the stochastic webs generated by kicked-oscillator Poincaré maps are indistinguishable from mathematical quasicrystals (Zaslavskii et al., 1991). For the patterns of periodic points (rather than the stochastic layers), one can in some cases prove the quasicrystalline symmetry rigorously (Lowenstein, 1993, 1994). Because we are specifically interested in the interplay between local and global behavior, we will restrict ourselves henceforth to the crystalline maps, and, in the interest of simplicity, to the case $\rho = 1/4$.

1.5 Normal and anomalous diffusive behavior

The local and global manifestations of chaos in the $\rho = 1/4$ stochastic web map have been studied numerically in great detail by Zaslavsky and Niyazov (1997), Zaslavsky et al. (1997), and Zaslavsky and Edelman (2001, 2003). That the long-time behavior is characterized by a diffusive power law (mean squared distance increasing as Dt) was verified for a wide range of parameter values. The empirical values of the effective diffusion constant D are shown in Fig. 1.6. The existence of sharp peaks in the plot suggested to the authors that perhaps the function $D(a)$ is actually singular for certain parameter values. In order to understand the possible breakdown of normal diffusive behavior in their data, the authors examined the Poincaré section in the fundamental cell for parameter $a = a_* = 6.349972$ chosen to be as close as possible to the most prominent apparent singularity. Figure 1.7(a) shows the phase portrait of the Poincaré map folded into the the square $[0, 2\pi)^2$. Although all but a small fraction of the square is occupied by a chaotic "sea" of orbit points, the distribution of points appears to be highly non-uniform. The orbit appears to have spent a great deal of time in the vicinity of four small island systems, leading to a prominent darkening of the picture in those locations. The effect comes into clearer focus by zooming in on the darker regions to reveal several levels of island chains. This is shown in Fig. 1.7(a—d). What distinguishes the parameter value a_* at which $D(a)$ blows up appears to be the self-similarity of the family of nested island chains.

If the diffusive behavior of the stochastic-web map breaks down at some $a = a_*$, what replaces it? The numerical experiments of Zaslavsky and Niyazov (1997) and Zaslavsky et al. (1997) provide an answer: for large times, the distance of an orbit point from the origin increases proportional to a super-diffusive power of the time:

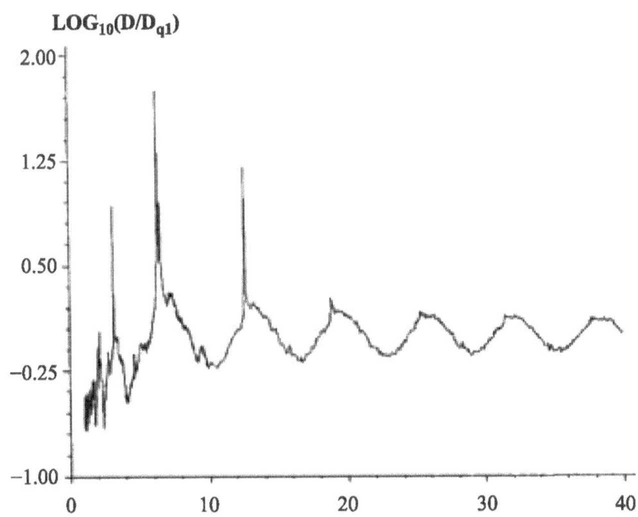

Fig. 1.6 Diffusion constant, divided by $D_{ql} = a^2/2$, as a function of parameter a for the 4-fold sinusoidal kicked-oscillator map, from Zaslavsky and Niyazov (1997), Fig. 2.

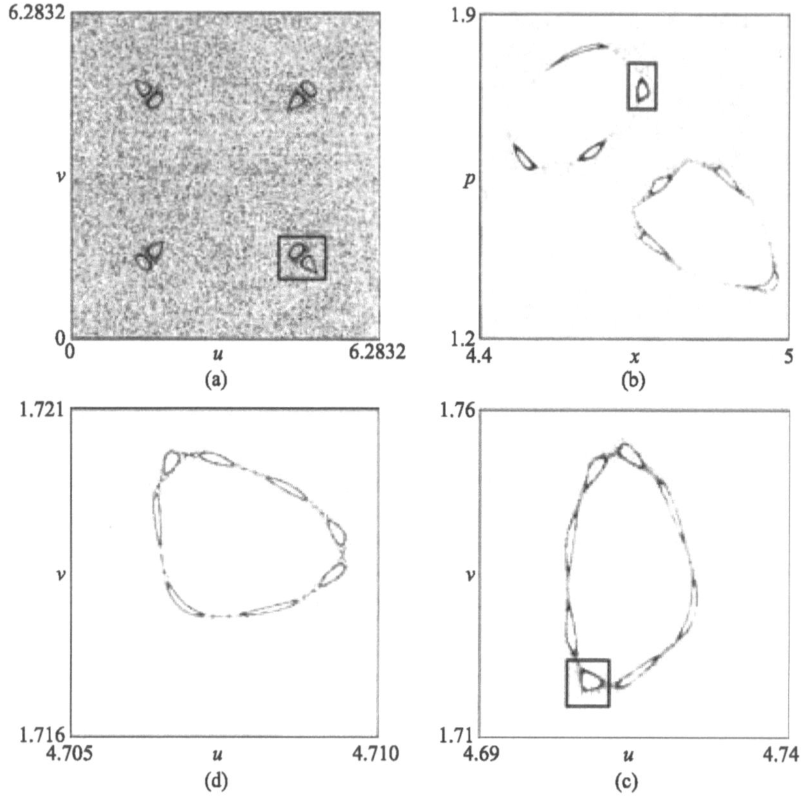

Fig. 1.7 Sticky chaotic orbit in the vicinity of one of the accelerator-mode islands for $a = a_*$, from Zaslavsky and Niyazov (1997), Figs. 3 and 5. In clockwise order, each frame is a zoom of the preceding one.

$|x| \sim t^{\mu}$, $\mu > 1/2$. The kind of process which produces such behavior is evident in Fig.1.8: globally, the chaotic orbit alternates between what appears to ordinary random-walk wandering and long, constant-velocity flights. The latter correspond to periods of "stickiness" in the vicinity of a self-similar island system, while the former correspond to excursions in the chaotic sea. The intermittent combination of the two produces an "anomalous" power law with an exponent intermediate between that of ballistic motion ($\mu = 1$) and a normal random walk ($\mu = 1/2$). In the numerical investigations of Zaslavsky and Niyazov (1997) and Zaslavsky et al. (1997), the relevant islands are accelerator modes: they are mapped from one cell to another by the action of the 4th-iterate map, and hence correspond to motion with constant average velocity. While trapped in the vicinity of such an island, a sticky orbit will temporarily share that velocity. Small accelerator-mode islands embedded in a chaotic sea are thus a good place to look for anomalous diffusive behavior. Dana (2004) has adopted such an approach to study a family of super-diffusive orbits in a $\rho = 1/3$ kicked oscillator model.

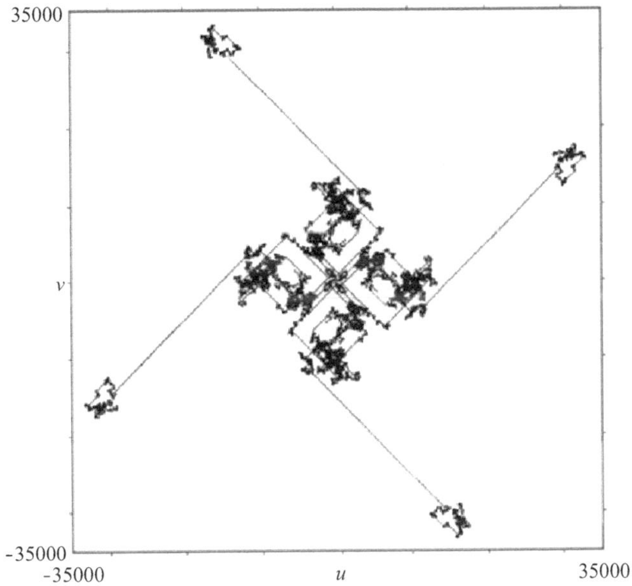

Fig. 1.8 Global orbits (2×10^6 iterations) for $a = 6.35$. Long ballistic flights correlated to long sojourns near self-similar island-around-island configurations, from Zaslavsky and Niyazov (1997), Fig. 1(a).

The discovery of power-law asymptotic behavior in numerical simulations of sinusoidal stochastic web maps with finely tuned parameter values has opened up a fascinating new realm to be explored both empirically and theoretically. Out of this research have emerged some promising and novel ideas concerning dynamical trapping by self-similar island structures, critical exponents for spatio-temporal scaling, and fractional-kinetic equations for describing anomalous transport, all under the rubric of "pseudochaos". For a comprehensive discussion of some of the main ideas in this rather broad field, the reader is referred to Zaslavsky (2005).

1.6 The sawtooth web map

In order to model pseudochaos with mathematical precision, one would like to consider relatively simple dynamical systems which exemplify in a distilled form those features which Zaslavsky and his collaborators found in the chaotic boundary regions of finely tuned web maps, namely (1) vanishing Lyapunov exponent and symbolic entropy, (2) self-similarity of the local dynamics, and (3) asymptotic diffusion like behavior of the global orbits, with a clearly delineated link between (2) and (3). Our choice for such a distilled model of pseudochaos is a particular class of 4-fold web maps with piecewise linear ("sawtooth") kick amplitudes (see Fig.1.9),

$$f(x) = \lambda(x - \lfloor x \rfloor_\tau \tau) \stackrel{\text{def}}{=} \lambda\{x\}_\tau, \qquad \lfloor x \rfloor_\tau \stackrel{\text{def}}{=} \lfloor x/\tau \rfloor \tau. \tag{1.8}$$

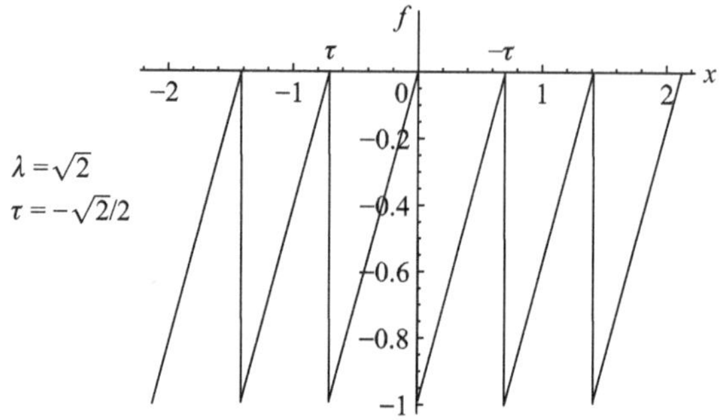

Fig. 1.9 Example of a sawtooth kick function $f(x)$.

Sawtooth kicked-oscillator models have been studied in some detail by Lowenstein et al. (2005), as an outgrowth of our work on piecewise isometries of convex polygons. We found that the piecewise isometric mappings of the square could be promoted to the local maps, on the fundamental cell, of sawtooth kicked-oscillator maps of the plane with rotation number $1/4$. Whereas Lowenstein et al. (2005) limit their attention to sawtooth maps with quadratic irrational parameters, the present investigation generalizes the "lifting" of piecewise isometries to include a (potentially) much larger class of models satisfying the so-called recursive tiling property.

The local map, being a piecewise linear mapping of the fundamental square Ω onto itself, is capable of generating fractal structures, thanks to the branching effect of its discontinuities, but lacks the stretching/contraction mechanism which generates chaotic layers and exponential divergence of nearby orbits in the presence of continuous kick functions.

1.7 Renormalizability

In rather general terms, *renormalizability* of a dynamical system refers to its *dynamical self-similarity*, i.e. the existence of an infinite hierarchy of scales, with geometrical and dynamical relations assuming the same form, apart from a scale transformation, at each level of the hierarchy (see, for example, MacKay, 1993).

In Chapter 2 of this book, we shall see how the concept of renormalizability is realized in the context of the sawtooth web map. There, the scaling hierarchy involves a nested sequence of convex polygonal domains, $D(L) \subset \Omega$, together with their respective *first return maps* $\rho(L)$ induced by K, $L = 0, 1, 2, \ldots$ The induced maps have the defining property that, for all $\mathbf{x} \in D(L)$, we have

$$\rho(L)(\mathbf{x}) = K^t(x),$$

where t, depending on x, is the smallest positive integer such that $K^t(x) \in D(L)$, but $K^s(x) \notin D(L)$ for all positive $s < t$.

The self-similarity of $D(L), \rho(L)$ pairs, for $L = 0, 1, 2, \ldots$, consists essentially of the recursion relations, valid for all levels L of the hierarchy:

(i) $\rho(L+1)$ is induced by $\rho(L)$ in a level-independent way.
(ii) $D(L+1) = S_\omega D(L)$, where S_ω is a scale transformation by a factor $\omega < 1$, with respect to some limit point $\mathbf{x}_\infty \in \Omega$.
(iii) $\rho(L+1) = S_\omega \circ \rho(L) \circ S_\omega^{-1}$.

The existence of a sequence of $D(L), \rho(L)$ pairs satisfying the above criteria (what we call a *scaling sequence*) is not always sufficient to ensure full renormalizability. Also needed is a criterion ensuring that the scaling sequence provides complete coverage of the aperiodic orbits of the model. For the sawtooth web maps, this criterion is *recursive tiling*: the return orbits of $\rho(L+1)$ should completely tile the region of $D(L)$ complementary to the periodic orbits of $\rho(L)$.

For most of the sawtooth web maps which we consider in detail in this book, renormalization indeed takes the form of a single scaling sequence of domains equipped with a return map hierarchy satisfying the recursive tiling property. All of these have parameter values with particularly simple properties: they are all in the interval $-2 < \lambda < 2$ and are irrational roots of quadratic polynomials $\lambda^2 + a\lambda + b$, $a, b \in \mathbb{Z}$. However these do not exhaust the cases of interest. In particular, we have cases where more than one scaling sequence is needed to completely cover the aperiodic orbits. Among the models with quadratic irrational λ, the case $\lambda = \sqrt{3}$ stands out: there are two disjoint scaling sequences, with distinct limit points and distinct temporal scaling properties, leading to fractal sets of aperiodic points with different dimensions. Our single example of a cubic irrational λ, the so-called $\pi/7$ model, has infinitely many disjoint scaling sequences, but even taking them together does not lead to complete coverage of the pseudochaotic web of aperiodic orbits.

What is needed for the $\pi/7$ model, and presumably more widely among non-quadratic models, is a renormalization scheme characterized by a catalogue of several polygons, each of which is recursively tiled by the return orbits of domains equivalent to the various catalogue members under translations, generalized rotations and reflections, and scale transformations. The hierarchy is then described by an infinite tree, rather than by a ladder or collection of ladders appropriate to quadratic models. The scheme is sketched in Fig. 2.10.

Obviously not all sawtooth web maps are renormalizable, even in the more general sense. The ones we have found are quite special, not only in the restriction to parameter values with special arithmetic properties, but also in the choice of the kick-function period τ and the placement of the fundamental domain Ω. In spite of what appears to be the highly restrictive nature of this class, we will find that there is considerable variety in the asymptotic behaviors on the phase plane which its members produce.

1.8 Long-time asymptotics

As we will see in Chapter 5 of this book, the hierarchical partition of phase space provided by recursive tiling allows us not only to prove general properties of asymptotically long orbits, but also to compute these orbits explicitly with great efficiency and exactitude. In particular, with the use of computer-manipulated algebraic numbers, the numerical errors inherent in floating point arithmetic can be completely avoided. Moreover, with recursive tiling and hierarchical symbolic dynamics, the time scales increase exponentially with level, making possible accelerated routes to the asymptotic regime.

Complementing the study of the long-time behavior of aperiodic orbits is a statistical formulation, embodied in the time evolution of the moments of a probability distribution. This is what we mean by *transport*. What does such a statistical formulation look like for our renormalizable kicked oscillator models? There are many possibilities, but, given the crystalline structure of the phase space, with τ-periodicity both horizontally and vertically on the x,y plane, as well as the local-global factorization of the dynamics, it is natural to adopt the following coarse-graining: average uniformly over local motions and define a probability distribution $P((m,n),t)$ solely for "walks" on the integer lattice $\{(m\tau,n\tau) : m,n \in \mathbb{Z}\}$.

In the definition of $P(\mathbf{r},t)$, the notion of random selection of initial conditions is the usual one of a uniform distribution with respect to some invariant measures. For the full set of orbits, Lebesgue measure (area) is the obvious choice. Since the full area measure of Ω is occupied by periodic polygons for the renormalizable models we are considering, it is clear that we could base our entire analysis on periodic orbits. This gives rise to the somewhat paradoxical situation that although the support of $P(\mathbf{r},t)$, i.e. the set of lattice points $\mathbf{r}$ where $P(\mathbf{r},t) > 0$, grows without bound with increasing t, the motion on any individual orbit with nonzero weight in the averaging is periodic, hence bounded.

In some interesting cases we will be primarily interested in the statistical long-time behavior of orbits living on specific invariant subsets of the exceptional set. Typically the latter occupies zero area, and so, in order properly to formulate the transport problem, it will be necessary to abandon Lebesgue measure. Hausdorff measure restricted to the particular sets of initial conditions will turn out to be a convenient alternative.

The picture which emerges from our analysis of long-time asymptotics is not quite as simple as one might have anticipated. The local self-similarity does not always lead to a global quasi-diffusivity: a number of renormalizable models have only uniformly bounded global orbits. Furthermore, where there is global expansion, it is not necessarily governed by the same power laws that one encounters locally. The main reason is that every iteration of the global map involves a 90° rotation of the plane, introducing a phase factor which must be included in calculations of the asymptotic scaling. The criteria for self-similarity are thus distinct from those which characterize the local scaling, and the associated power laws have different exponents.

1.9 Linking local and global behavior

What makes the study of the sawtooth kicked-oscillator maps so compelling is the fact that the local renormalizability and global transport can be tied together in a single, satisfying package. Central to this linkage is the fact that the local map K, a piecewise linear map on the square, and the global map L, a lattice isometry, are both governed by a single abstract map acting on symbol strings. This is not the simple "coin-toss" coding based on the elementary partition of Ω, but rather a hierarchical one reflecting the level structure of recursive tiling. In such a coding, which we will discuss fully in Chapter 3, successive symbols refer to successive levels in the renormalization hierarchy, and hence compress space and time at a geometric rate as one proceeds along the symbol string. The dynamical map on the symbol strings is of a lexicographic type developed by Vershik (1985), with applications in diverse mathematical contexts. In our numerical explorations of the long-time regime, the updating rule for symbol strings acts as an abstract machine which we call a *codometer* (see Appendix B). It is reminiscent of, but more general than, that of an ideal odometer with infinitely many registers, and will allow highly efficient computational excursions into the long-time regime.

For the quadratic models, the story does not end here. As we shall see in Chapter 7, the symbolic dynamics of the kicked-oscillator models also underlies the lattice dynamics of certain models of non-dissipative round-off. In spite of the common abstract structure, the lattice transformations are vastly different from the kicked-oscillator lattice maps, and are endowed with completely different long-time asymptotic behavior. By means of a substitution-based analysis of the orbits in the long-time regime, we succeed in deriving, in one of our models, a rigorous expression for the asymptotic power law with log-periodic modulation. By means of highly efficient, codometer driven programs, we have verified the general form of the modulated power laws in other globally expansive models as well, including one where the modulating factor for the second moment is log-quasiperiodic.

1.10 Organization of the book

The remainder of the book is organized as follows.

Chapter 2

We begin with a review of the concepts and formalism of scaling, renormalization, and recursive tiling, all within the context of the local map. This includes a generalization of the renormalization scheme to cover a wider class of models, including the so-called $\pi/7$ model with its cubic irrational parameter.

Chapter 3

We establish the correspondence of hierarchical symbol sequences with points on the discontinuity-avoiding aperiodic orbits. After introducing the dynamical map, in the form of an odometer-like updating scheme, the codometer, we concentrate on the technically complicated problem of deriving intrinsic rules for the admissibility of symbol sequences.

Chapter 4

Here we focus on the measure-theoretic and fractal (or multifractal) properties of the residual set of the local map. The introduction of Hausdorff dimension and Hausdorff measure will be essential, later on, to formulating transport on interesting sets of Lebesgue measure zero. Using a transfer matrix to link successive levels of the renormalization hierarchy, we calculate a nontrivial spectrum of generalized dimensions for the $\pi/7$ model.

Chapter 5

We now examine how the local return maps of a scaling sequence can be lifted to the infinite kicked-oscillator phase space. In some cases the resulting global translation vectors are *globally expansive*, increasing in length geometrically with increasing level. In such cases, we prove that the discontinuity-avoiding aperiodic orbits tend to infinity according to a power law. We study this phenomenon numerically in a number of examples. In the particularly interesting case of the $\pi/7$ model, we see that different initial conditions can lead to a variety of asymptotic power laws.

Chapter 6

A statistical formulation of the time evolution is presented in this chapter, based on a coarse-grained probability distribution $P((m,n),t)$. Once again, the renormalizability of the models and especially the recursive tiling property, serve as the organizing principles. We consider examples in which the probability is supported on the 2-dimensional periodic domains, but also ones restricted to interesting lower-dimensional sets. Highly efficient codometer updating allows us to investigate numerically asymptotic power laws with log-periodic modulation.

Chapter 7

In this final chapter, we consider a different type of kicked oscillator dynamics, namely that of a Hamiltonian round-off map (Lowenstein et al., 1997; Lowenstein and Vivaldi, 1998, 2000). Once again we have a local piecewise isometry which induces a permutation of $\mathbb{Z}^2$. Although in this class of models all orbits are periodic, one can still describe the long-time behavior statistically in terms of periodically and quasi-periodically modulated power laws. We study these both analytically and with highly efficient numerical experiments.

Appendixes

For the sake of completeness, we include in the Appendix A a number of data tables for the models which appear throughout the book. In addition, we have a short explanation of our implementation of the hierarchical symbolic dynamics, the codometer, in Appendix B.

References

Dana I. (2004) Global superdiffusion of weak chaos, *Physical Review E* **69**, 016212.

Lowenstein J.H. (1993) Quasiperiodic structure of the stochastic web map, *Physical Review E* **47** (Rapid Communication), 3811–3814.

Lowenstein J.H. (1994) Quasilattice of fixed points of the fivefold stochastic web map, *Physical Review E* **49**, 232–246.

Lowenstein J.H. and Vivaldi F. (1998) Anomalous transport in a model of Hamiltonian roundoff, *Nonlinearity* **11**, 1321–1350.

Lowenstein J.H. and Vivaldi F. (2000) Embedding dynamics for round-off errors near a periodic orbit, *Chaos* **10**, 747–755.

Lowenstein J.H., Hatjispyros S., and Vivaldi, F. (1997) Quasi-periodicity, global stability and scaling in a model of Hamiltonian round-off, *Chaos* **7**, 49–66.

Lowenstein J.H., Poggiaspalla G., and Vivaldi F. (2005) Sticky orbits in a kicked-oscillator model, *Dynamical Systems* **20**, 413–451.

MacKay R.S. (1993) *Renormalisation in Area-Preserving Maps*, World Scientific, Singapore.

Vershik A.M. (1985) A theorem on the Markov periodical approximation in ergodic theory, *Journal of Soviet Mathematics* **28**, 667–674.

Zaslavskii G.M., Zakharov M. Yu., Sagdeev R.Z., Usikov D.A., and Chernikov A. A. (1986) Generation of ordered structures with a symmetry axis from a Hamiltonian dynamics, *JETP Letters* **44**, 451–456.

Zaslavskii G.M., Sagdeev R.Z., Usikov D.A., and Chernikov A. A. (1991) *Weak Chaos and Quasiregular Patterns*, Cambridge University Press, Cambridge.

Zaslavsky G.M. (2005) *Hamiltonian Chaos and Fractional Dynamics*, Oxford University Press, Oxford.

Zaslavsky G.M. and Edelman M. (2001) Weak mixing and anomalous kinetics along filamented surfaces, *Chaos* **11**, 295–305.

Zaslavsky G.M. and Edelman M. (2003) Pseudochaos, *Perspectives and Problems in Nonlinear Science: A Celebratory Volume in Honor of Lawrence Sirovich*, eds. E. Kaplan, J. Marsden, and K. R. Sreenivasan, Springer, New York, 421–423.

Zaslavsky G.M. and Niyazov B.A. (1997) Fractional kinetics and accelerator modes, *Physics Reports* **283**, 73–93.

Zaslavsky G.M., Edelman M., and Niyazov B.A. (1997) Self-similarity, renormalization, and phase space nonuniformity of Hamiltonian chaotic dynamics, *Chaos* **7**, 159–181.

Chapter 2
Renormalizability of the Local Map

The local component of the sawtooth kicked-oscillator dynamics is the renormalizable, piecewise linear mapping K of a square Ω onto itself,

$$K : \Omega \to \Omega, \qquad K(\mathbf{u}) = \mathbf{C} \cdot \mathbf{u} \bmod \Omega, \tag{2.1}$$

where

$$\mathbf{C} = \begin{pmatrix} \lambda & 1 \\ -1 & 0 \end{pmatrix}, \qquad \Omega = \begin{cases} [0,\tau)^2 & \tau > 0, \\ (\tau,0]^2 & \tau < 0. \end{cases}$$

Here the "mod" operation is a lattice translation to enforce $K(\mathbf{u}) \in \Omega$. The current chapter is a thorough examination of the map K for an important class of renormalizable models.

The piecewise affine map K on the square has an important place in the dynamical systems literature of the last two decades. Some of the interesting examples are:

- Overflow oscillations of a second-order nonlinear digital filter: Ashwin et al. (1997), Chua and Lin (1990), Davies (1995), Wu and Chua (1993).
- Equivalence to the sawtooth analogue of the Chirikov-Taylor standard map: Ashwin (1997), Dana (2004).
- Sticky orbits in stochastic-web maps: Zaslavsky and Edelman (2003), Lyubomudrov et al. (2003).
- Localization of the lattice orbits of oscillator models with Hamiltonian round-off: Lowenstein et al. (1997), Lowenstein and Vivaldi (2000).
- Piecewise isometries of convex polygons: Adler et al. (2001), Akiyama et al. (2008), Buzzi (2001),Goetz (1996, 1998, 2000, 2001), Goetz and Poggiaspalla (2004), Kahng (2000, 2002), Kouptsov et al. (2002), Lowenstein (2007), Lowenstein et al. (2004), Poggiaspalla (2006).

The material of the current chapter is drawn in large part from Kouptsov et al. (2002) and Lowenstein et al. (2004), as well as the electronic supplements linked to those articles, to which the interested reader is directed for additional details. For the sake of notational consistency with these articles, we introduce a trivial change of phase-space coordinates at this point, namely a canonical transformation which essentially interchanges the roles of position and momentum, namely

$$\begin{pmatrix} x \\ y \end{pmatrix} \mapsto \begin{pmatrix} y \\ -x \end{pmatrix}.$$

Adopting the transformed canonical frame through the remainder iof the book, we redefine the web map W of (1.4) and matrix $\mathbf{C}$ of (2.1) to be

$$W\begin{pmatrix} x \\ y \end{pmatrix} = \begin{pmatrix} y \\ -x+f(y) \end{pmatrix}, \qquad \mathbf{C} = \begin{pmatrix} 0 & 1 \\ -1 & \lambda \end{pmatrix}.$$

2.1 Heuristic approach to renormalizability

The restrictions imposed by renormalizability on the parameter λ and square domain Ω in (2.1) are obviously quite severe. Currently, no theorem is available to tell us which λ are allowed, and we will not make an attempt to explore this problem. Our aim will be a more practical one, namely to construct a class of renormalizable local maps associated with an interesting and diverse family of kicked oscillator models. In each case, we demand that the asymptotic long-time behavior be determined with mathematical precision. Our choice of models will thus be governed by heuristic considerations, while our treatment of each model will proceed without sacrifice of rigor.

2.1.1 Generalized rotations

The matrix $\mathbf{C}$ has eigenvalues

$$\frac{1}{2}(\lambda \pm \sqrt{\lambda^2-4}).$$

which, for $|\lambda| < 2$, are complex conjugate points $e^{\pm i\theta}$ on the unit circle. The matrix is thus linearly conjugate to a planar rotation by θ,

$$\mathbf{R}(\theta) = \begin{pmatrix} \cos\theta & \sin\theta \\ -\sin\theta & \cos\theta \end{pmatrix}$$

and the mapping K is linearly conjugate to a piecewise isometry on the rhombus.

If we assign to the original xy plane the metric tensor

$$\mathscr{Q} = \begin{pmatrix} 1 & -\frac{\lambda}{2} \\ -\frac{\lambda}{2} & 1 \end{pmatrix} \tag{2.2}$$

we can legitimately refer to **C** as a rotation and K itself as a piecewise isometry (PWI for short). We shall routinely use this terminology in what follows.

At first sight we cannot rule out $|\lambda| \geq 2$, for which the eigenvalues of **C** are real. However, for reasons which will soon become clear, a rotational **C** turns out to be essential to our strategy of constructing renormalizable models, and so henceforth we shall assume $|\lambda| < 2$.

2.1.2 *Natural return map tree*

Renormalizability entails self-similar families of polygonal domains, but given only the map K, how does one find those subdomains of Ω which have the desired structure? The only successful strategy we have discovered (and even then, only for carefully selected parameter values) is a recursive search of the *natural return map tree*. Let us describe informally how this works.

The natural return map tree takes as its starting point (level 0) the partition of Ω into a convex polygonal domains (known as atoms)

$$\Omega_i = \{z \in \Omega : \iota(z) = i\}, \quad i = 0, \ldots, a-1.$$

We then calculate the (first-) return map ρ_i on each domain Ω_i. This is defined, for almost all $z \in \Omega_i$, by

$$\rho_i(z) = K^t(z),$$

where the return time t is the smallest positive integer such that $K^t(z) \in \Omega_i$. A constructive algorithm for ρ_i, which emphasizes its role as a piecewise isometry on Ω_i, is the following:

Induction algorithm. Start with a specific Ω_i. Apply K and intersect the result with each of the a atoms Ω_j to obtain

$$E_{i,j} = K(\Omega_i) \cap \Omega_j.$$

If $E_{i,i} \neq \emptyset$, we note that a subdomain of Ω_i, namely

$$D_i = K^{-1}(E_{i,i})$$

has returned to Ω_i after one iteration of K, and hence qualifies as a member of the set $\mathscr{A}(\rho_i)$ of atoms of the (level 1) return map ρ_i of Ω_i. The remaining nonempty $E_{i,j}$, with $j \neq i$, constitute a set of "in transit" domains which need to be subjected to additional iterations of K to bring them home. In the next step, we calculate

$$E_{i,j,k} = K(E_{i,j}) \cap \Omega_k,$$

appending the nonempty $D_{i,j} = K^{-1}(E_{i,j,i})$ to the list of atoms $\mathscr{A}(\rho_i)$ and designating the remaining nonempty $E_{i,j,k},\ j,k \neq i$, as candidates for further mapping and slicing.

Continuing as above, we obtain a recursive construction of a set of atoms

$$D_{i,i_1,i_2,\ldots,i_{n-1}}, \quad i_1,\ldots,i_{n-1} \neq i.$$

For each n, these are convex subdomains of Ω_i which return to that polygon after a *return time* of n, along the *return paths*

$$(i,i_1,\ldots,i_{n-1}).$$

Specifically, the return time denotes the number of K iterations needed for first return, and the return path lists the atoms of the level-0 map visited along the way. Note that we have no guarantee that the construction ever terminates, and so the number of atoms of the induced map ρ_i may be countably infinite.

It is clear that there are no formal differences between the original piecewise isometry K and the return maps ρ_i induced via the above construction on the atoms of K. It is thus completely straightforward to apply the analogous construction recursively to obtain a tree in which each node of level L is a *dressed domain*, i.e. a pair $\Delta = (D,\rho)$ where D is a convex polygon (perhaps including some or all of its edges) and ρ is the (first-) return map induced by the level-$(L-1)$ dressed domain containing Δ. Clearly, level 0 of the tree consists of the dressed domain (Ω,K), while level 1 consists of $\{(\Omega_i,\rho_i) : i = 0,\ldots,a-1\}$.

In terms of the natural return map tree, the search for renormalizability reduces to an attempt to find two dressed domains, $\Delta(0) = (D(0),\rho(0))$ and $\Delta(1) = (D(1),\rho(1))$, the latter a direct descendant of the former, differing from it by a similarity transformation S, i.e.

$$D(1) = S(D(0)), \qquad \rho(1) = S \circ \rho(0) \circ S^{-1}.$$

Most simply, S will be a scale transformation by a factor $\omega < 1$, but could also involve other linear transformations which preserve the $\mathscr{Q}$-metric. Having found one such pair on the tree, we are guaranteed of an infinite sequence, namely

$$D(L) = S^L(D(0)), \qquad \rho(L) = S^L \circ \rho(0) \circ S^{-L}.$$

In what follows we will refer to $\{\Delta(L) : L = 0,1,2,\ldots\}$ as a *scaling sequence* and $\Delta(0)$ as a *scaling domain*.

Suppose for a particular choice of (K,Ω) we have been fortunate enough to find a scaling sequence of dressed domains buried in the natural return map tree. Does the model qualify as a *renormalizable* one? The answer will be in the affirmative if the scaling sequence in some way accounts for all of the essential dynamics of the model. The criterion which we adopt to make this concept precise is *recursive tiling*: the return map orbits of $D(0)$ should tile all of Ω, apart from a finite number of periodic K-orbits, and the return map orbits of each $D(L)$, $L > 0$, should tile all of $D(L-1)$, again up to finitely many periodic $\rho(L-1)$-orbits. We will also want to consider renormalizable those models with a finite number of disjoint scaling sequences which together tile the entire square, again up to finitely many periodic orbits.

2.1.3 Examples

To illustrate the return map construction and quest for renormalizable models described in abstract terms above, it is useful to study in detail two explicit examples. The first of these has the deceptively simple choice $\lambda = 1/2$, with K defined on the unit square $\Omega = [0,1)^2$. The map has two atoms, Ω_0 and Ω_1. The first 8 steps in the construction of the return map partition of the triangular domain Ω_0 are shown in Fig. 2.1.

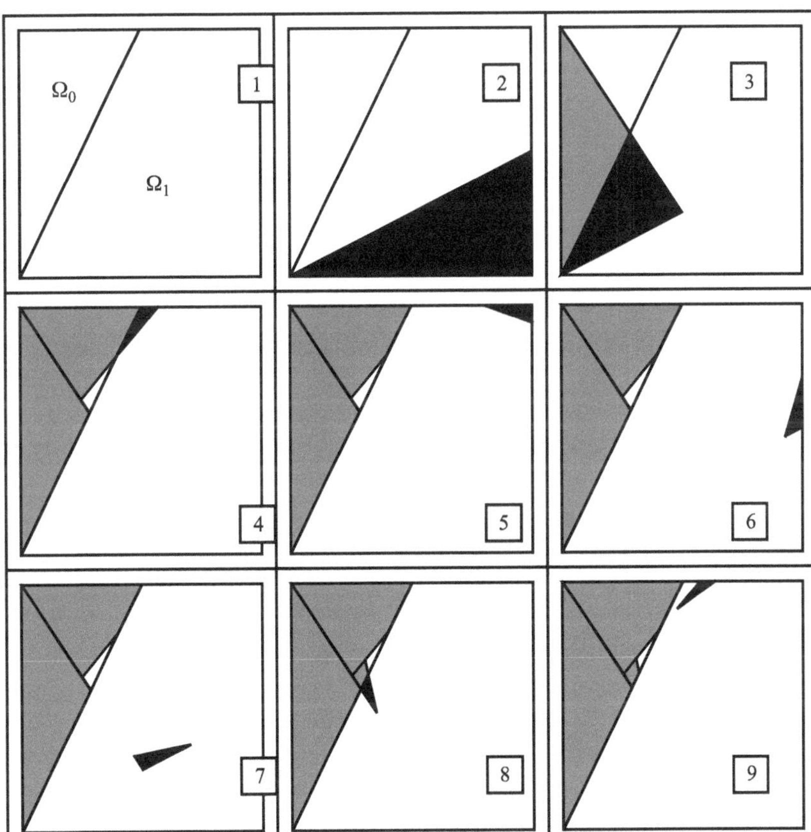

Fig. 2.1 First 8 steps of the construction of the return map ρ_0 on Ω_0.

After two applications of K, the image triangle overlaps the two atoms and is split into $E_{0,0} \subset \Omega_0$ (gray) and $E_{0,1} \subset \Omega_1$ (black). The former, mapped back with K^{-1}, gives us the first of the atoms of ρ_0, the induced return map on Ω_0. The latter will require further mapping and slicing to produce the remaining members of $\mathscr{A}(\rho_0)$. Another branching event can be seen in the eighth panel. Obviously the construction does not halt here, and continuing the process for a large number of iterations does not complete the job either. In fact, it is possible that it will never halt, i.e. the return map partition may be infinite.

The return map for Ω_1 is dramatically simpler (the construction terminates after two steps), but when we proceed to the second level of the return map tree, we find that the construction of the return map partition of each of the atoms does not appear to terminate after many iterations.

From the point of view of a reasonable search procedure, the model with $\lambda = 1/2$, $\Omega = [0,1)^2$ appears to be a most unpromising candidate. The reason is not difficult to understand. Consider the last step of a return map construction for a domain Ξ which actually terminates. Here we have a convex polygon D outside of Ξ (colored black in the figures) and a convex polygon E within Ξ (colored white in the figures), such that $E = K(D)$. A necessary condition for successful termination is clearly that the edges of polygons D and E should have the same relative orientations, with the orientations of E obtained from those of D by a single rotation C. This can be checked quite easily, since every domain boundary, by construction, has the orientation of a line $C^m \cdot \phi$, where m is a nonnegative integer and ϕ is the line through $(0,0)$ and $(1/2,1)$. The difficulty in achieving termination can be seen in frames 2, 3, and 7 of Fig. 2.1, where the black and white triangles have a "bow-tie" configuration, with orientations $(C^{m_1}\phi, C^{m_2}\phi, C^{m_3}\phi)$ and $(C^{m_2}\phi, C^{m_1}\phi, C^{m_4}\phi)$, with m_1, m_2, m_3, m_4 all different. Since $(m_1+1, m_2+1, m_3+1) \notin \{(m_2,m_1,m_4),(m_1,m_4,m_2),(m_4,m_2,m_1)\}$, the corresponding return orbits cannot terminate the construction.

The above discussion suggests a strategy which might enhance our chances of finding renormalizable models by means of a systematic search of the natural return-map tree. Suppose C^q were equal to the identity for some small value of q, corresponding to a *rational rotation number* p/q. There would then be only a small number q of possible orientations of lines, which would greatly increase the chances for finite (perhaps even small) return map partitions, as well as for finding similar polygons among the nodes of the tree. It is not difficult to show that the (necessarily irrational) parameter $\lambda = 2\cos(2\pi p/q)$ is an *algebraic integer of degree n*, i.e. the solution of a polynomial equation of the form

$$\lambda^n = c_0 + c_1\lambda + \cdots + c_{n-1}\lambda^{n-1}, \tag{2.3}$$

where $n, c_k \in \mathbb{Z}$, $n > 1$. For small values of n, only a small number of possible rotation numbers are allowed. Thanks to (2.3), regarded as a recursion relation for powers of λ, the set

$$\mathbb{Q}(\lambda) = \{a_0 + a_1\lambda + \cdots + a_{n-1}\lambda^{n-1} : a_i \in \mathbb{Q}, i = 0, \ldots, n-1\},$$

is an *algebraic number field*, closed with respect to all the operations of ordinary arithmetic, including division by non-zero elements. Thus, provided that the kick-function period τ belongs to $\mathbb{Q}(\lambda)$, all of the coordinates of polygonal vertices arising in our return map construction will also be restricted to $\mathbb{Q}(\lambda)$.

It is not immediately obvious that choosing λ to be a low-degree algebraic integer should help our search for dynamical self-similarity (beyond the restriction that it places on the denominator of the rotation number). Of course, it is well known that the lowest-degree algebraic integers, solutions of quadratic equations, enjoy algebraic self-similarity in their continued fraction expansions. Moreover, for one-dimensional maps analogous to piecewise isometries, namely the interval exchange transformations, one has a powerful theorem of Boshernitzan and Carroll (1997) establishing their renormalizability for quadratic irrational parameters. Unfortunately, no comparable theorem for two-dimensional PWI's has been proved. However, for two-dimensional PWI's, the renormalizability of an important class of models with quadratic irrational λ has been rigorously established by Kouptsov et al. (2002) using computer assisted proofs. It is here that the true advantage of the restriction to low-degree algebraic numbers makes itself felt: it makes it possible to use computer software to perform exact calculations on specific models, most of which have exceedingly complicated multi-level return map structures, thereby verifying important properties of each model and, by exhaustion, the entire class.

Before examining three particularly interesting models from the class of PWI's of the square with rational rotation numbers and quadratic irrational parameters, it will be useful to illustrate how the systematic search for renormalizable return map structure succeeds in a particularly simple example. The contrast with the $\lambda = 1/2$ case will be striking.

We choose a model with rotation number $1/5$, namely

$$\lambda = 2\cos(2\pi/5) = \frac{1}{2}(\sqrt{5}-1), \qquad \Omega = (\tau, 0]^2, \qquad \tau = -\lambda.$$

For simplicity, we ignore here the orbits of points on the boundaries and line of discontinuity. The atoms may then be considered as open polygons specified by their respective lists of vertices, in square brackets:

$$\Omega_0 = \left[(0,0),(0,\tau),(-\tau^2,\tau)\right], \qquad \Omega_1 = \left[(0,0),(-\tau^2,\tau),(\tau,\tau),(\tau,0)\right].$$

The partition is shown in frame 1 of Fig. 2.2.

We now explore a pathway on the return map tree which will reveal a scaling domain at level 2. We first construct the return map for Ω_1. The 2-step process is shown in Fig. 2.2. Of the two level-1 atoms, we choose to construct the return map for the triangular one. Again, the construction requires only 2 steps to terminate, as shown in Fig. 2.3. We are now at level 2. Next, we construct the return map of the right-hand triangle. This time 3 steps are required, as shown in Fig. 2.4. Now we have reached level 3. Finally, we compute, in 3 steps, the level 4 return map of the upper triangular atom, as shown in Fig. 2.5. Comparing the level 2 and level 4 dressed domains in Figs. 2.3 and 2.5, we see that they appear to be identical, up to a rescaling. Explicit calculation confirms this and gives the scale factor as $\omega = 1 + \tau = \frac{1}{2}(3 - \sqrt{5})$.

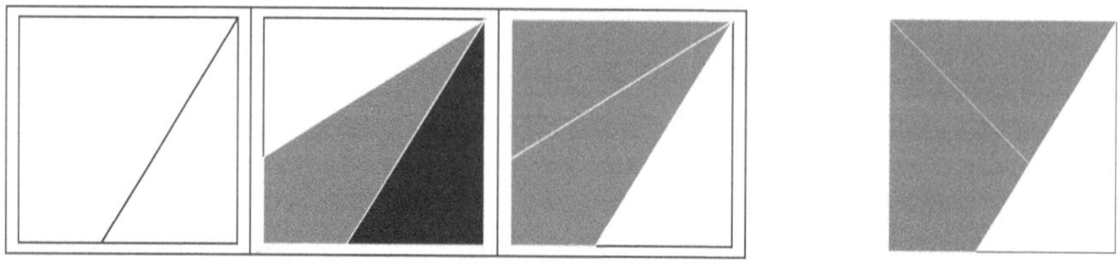

(a) (b)

Fig. 2.2 **(a)** Construction of the return map for Ω_1. **(b)** Atoms of the level 1 return map.

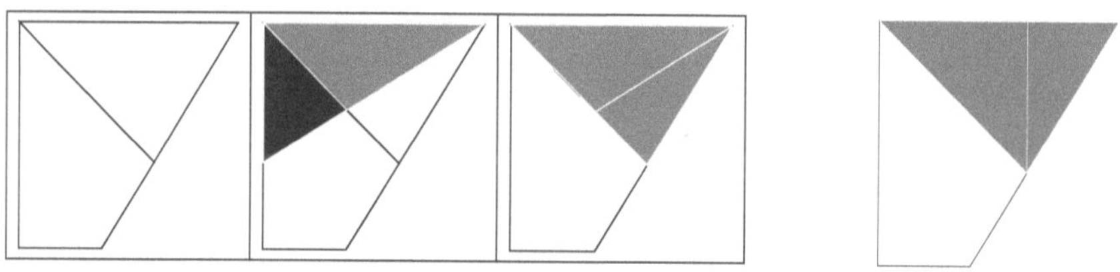

(a) (b)

Fig. 2.3 **(a)** Construction of the return map for the triangular level 1 atom. **(b)** Atoms of the level 2 return map (gray).

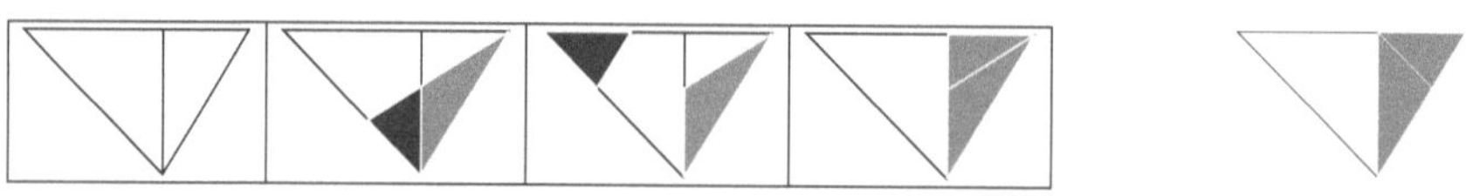

(a) (b)

Fig. 2.4 **(a)** Construction of the return map for the right-hand level 2 atom. **(b)** Atoms of the level 3 return map (gray).

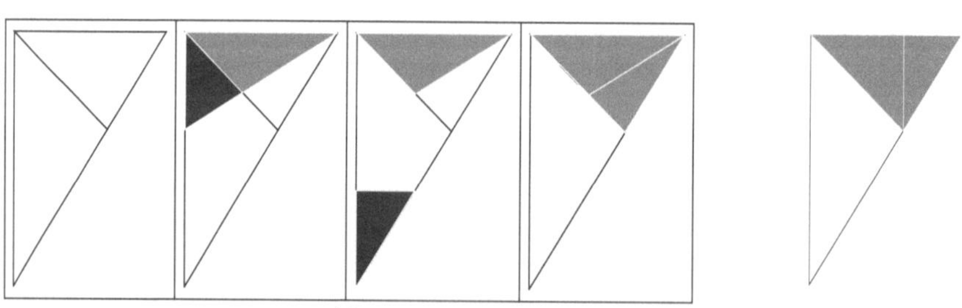

(a) (b)

Fig. 2.5 **(a)** Construction of the return map for the upper level 3 atom. **(b)** Atoms of the level 4 return map (gray).

As a final step in checking the renormalizability of this simple model, let us verify the recursive tiling property, both for the level 2 dressed domain $\Delta(0) = (D(0), \rho(0))$ and its level 4 granddaughter, $\Delta(1) = (D(1), \rho(1))$. The return K-orbits of the atoms of $\Delta(0)$ are seen in Fig. 2.6(a) to tile all of Ω, with the exception of a pentagon which is mapped into itself by K. Similarly, the $\Delta(0)$ orbits of the atoms of $\Delta(1)$ are seen in Fig. 2.6(b) to tile all of $D(0)$, apart from two pentagons which are mapped into themselves by $\rho(0)$. Combining the two tilings, we display the K-orbits of the atoms of $\Delta(1)$ in Fig. 2.6(c). Again the coverage is complete apart from the periodic pentagons.

By going further up the scaling sequence, we reveal more and more of the self-similar return map structure, as well as more and more of the periodic domains of the model. To see this, we plot in Fig. 2.7 the $\Delta(0)$ orbits of the atoms of $\Delta(4) = (D(4), \rho(4))$, where $D(4)$ is $D(0)$ rescaled by a factor ω^4, and the $\Delta(0)$ return paths are obtained by iterating 4 times the substitutions

$$0 \mapsto 1,0,0, \qquad 1 \mapsto 1,0,0,1,1.$$

(a) (b) (c)

Fig. 2.6 **(a)** Tiling of Ω by atoms of Δ_1. **(b)** Tiling of D_1 by atoms of Δ_2. **(c)** Tiling of Ω by atoms of Δ_2.

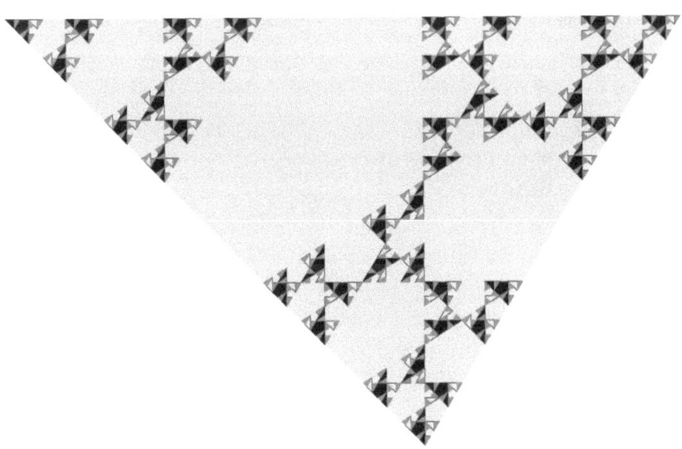

Fig. 2.7 Tiling of $\Delta(0)$ by atoms of $\Delta(4)$.

We can calculate from the return paths the total return times for the $\Delta(4)$ atoms, namely 171 and 341, approximately 4 times the corresponding return times for $\Delta(3)$, and 16 times those of $\Delta(2)$. It appears that for $n \to \infty$, the entire area occupied by the periodic pentagons of various sizes is tending toward the full area of the triangle $D(0)$, while the complementary *exceptional set*, covered by the $\Delta(L)$ return orbits, is tending toward a zero-area fractal. In fact the recursive process constructing the exceptional set via a sequence of coverings is analogous to that of a two-dimensional Cantor set, e.g. the well-known Sierpinski gasket, and its dimension can readily be estimated. Since the diameters of the covering triangles are decreasing uniformly with L as ω^L, while the number N_L of covering triangles can be seen to be increasing roughly as 4^L, the resulting fractal dimension d, defined for such Cantor set constructions (Falconer, 1990) by

$$N_L \sim 4^L \sim (\omega^L)^{-d},$$

gives

$$d = -\frac{\log 4}{\log \omega} = 1.44\ldots.$$

2.2 Quadratic piecewise isometries

We now treat systematically a family of piecewise isometries K for which renormalization can be carried out exactly using algebraic numbers and integer arithmetic. The family consists of maps (2.1) with the following properties:

(i) The rotation number is a rational number p/q.
(ii) The parameter $\lambda = 2\cos 2\pi p/q$ is a root of a polynomial $x^2 + bx + c$, $b, c \in \mathbb{Z}$. We call the conjugate root $\tilde{\lambda}$.
(iii) The square Ω is $[0,\tau)^2$ (resp. $(\tau,0]^2$) for positive (resp. negative) $\tau = \tilde{\lambda}^{-1}$.
(iv) The point $(0,0)$ is a fixed point.

The third and fourth defining criteria may seem, at this point, somewhat arbitrary. They arose from the role of this particular family in models of Hamiltonian round-off, a subject which we shall revisit in Chap. 7. In any case, they provide a convenient set of constraints on the otherwise unmanageably large class of piecewise isometries satisfying the first two criteria.

2.2.1 Arithmetic preliminaries

The parameter λ is the trace of the matrix C in (2.1), and hence

$$\lambda = \xi + \xi^{-1}, \qquad \xi = e^{2\pi i p/q},$$

where, without loss of generality, we may assume that p and q are relatively prime positive integers, i.e. their greatest common divisor is 1. The fact that λ satisfies a quadratic polynomial equation implies that ξ satisfies a quartic equation. From number theory (Hardy and Wright, 1979), one learns that the $\phi(q)$ roots of unity $e^{2\pi ip/q}$ with p running over the set of positive integers with $\gcd(p,q) = 1$ are roots of the equation

$$H_q(\xi) = 0,$$

where H_q is the degree-$\phi(q)$ *cyclotomic polynomial*

$$H_q(\xi) = \prod_{d|q}(\xi^d - 1)^{\mu(q/d)}.$$

Here $\phi(q)$, the *Euler totient function*, is the number of positive integers not exceeding q which are coprime to q. Moreover, $\mu(n)$, the *Moebius function*, is equal to 1 if $n = 1$, 0 if n has a square divisor greater than 1, and $(-1)^k$ if n is not divisible by any square greater than 1 and has k prime divisors. For $q > 2$, one has

$$H_q(x) = G_q(x + x^{-1})x^{-\phi(q)/2}.$$

For the specific case $\phi(q) = 4$, one can show (Kouptsov et al., 2002) that

$$H_q(x) = x^4 - \mu(q)x^3 + h(q)x^2 - \mu(q)x + 1,$$

and

$$G_q(x) = x^2 - \mu(q)x + h(q) - 2, \tag{2.4}$$

where, for $\phi(q) \geq 4$,

$$h(q) = \begin{cases} \mu(q)(\mu(q) - 1)/2 & q \text{ odd}, \\ \mu(q)(\mu(q) + 1)/2 & 4 \nmid q, \\ -\mu(q/2) & 4|q. \end{cases}$$

As a consequence of $G_q(\lambda) = G_q(\tilde{\lambda}) = 0$, we have

$$\tilde{\lambda} = \mu(q) - \lambda.$$

The possible values q consistent with quadratic G_q are, from the preceding, those for which $\phi(q) = 4$, namely 5,8,10,12. In each case there will be four p values such that $\gcd(p,q) = 1$. Since numerators p and $q - p$ lead to the same value of $\lambda = 2\cos(2\pi p/q)$, the choices of rotation number reduce to the following eight:

$$\frac{p}{q} \in \left\{\frac{1}{5}, \frac{2}{5}, \frac{1}{8}, \frac{3}{8}, \frac{1}{10}, \frac{3}{10}, \frac{1}{12}, \frac{5}{12}\right\}.$$

As the root of a polynomial of the form $x^d + c_{d-1}x^{d-1} + \cdots + c_0$, with d and all coefficients integers, the parameter λ is an *algebraic integer*. In what follows, several sets of algebraic numbers associated with λ will play an important role, namely the *ring* $\mathbb{Z}[\lambda]$, the *field* $\mathbb{Q}(\lambda)$, and the $\mathbb{Z}[\lambda]$-*module* $\mathcal{M}(\tau)$ defined by

$$\begin{aligned}\mathbb{Z}[\lambda] &= \{m+n\lambda : m,n\in\mathbb{Z}\},\\ \mathbb{Q}(\lambda) &= \{r+s\lambda : m,n\in\mathbb{Q}\},\\ \mathcal{M}(\tau) &= \tau\mathbb{Z}[\lambda] = \{m+n\tau : m,n\in\mathbb{Z}\},\end{aligned}$$

The last of these will be of particular importance in specifying the boundaries of domains in return map partitions, exploiting the property of closure of the module under multiplication by elements of $\mathbb{Z}[\lambda]$.

We note that the principal difference between $\mathbb{Z}[\lambda]$, a ring, and $\mathbb{Q}(\lambda)$, a field, is the fact that elements of the former are not required to have multiplicative inverses in $\mathbb{Z}[\lambda]$, and so the nonzero elements do not form a group. Those elements which do have inverses are called *units*, and can be shown to be roots of polynomials $x^d + c_{d-1}x^{d-1} + \cdots + c_0$ for which $c_0 = \pm 1$. For example, in the ring $\mathbb{Z}[\sqrt{2}]$, the algebraic integer $\sqrt{2}+1$ is a unit, since its inverse, $\sqrt{2}-1$, is also in the ring. One checks that $\sqrt{2}+1$ is a root of the polynomial x^2-2x-1.

In each of the models we will study with quadratic irrational λ, all units, including the scale factor appearing in its renormalization, will be of the form $\pm\eta^n$, $n\in\mathbb{Z}$, with a unique *fundamental unit* η. In contrast, the units in the cubic model which we will introduce later in this chapter are generated by two independent fundamental units, consistent with the more intricate renormalization which takes place in that model.

2.2.2 Domains

For the piecewise isometry K with rational rotation number p/q, the discontinuity set Γ consists of all forward and backward images of the intersection of the x axis with the square Ω. Given that $\mathbf{C}^q$ is the identity matrix, it is clear that the line segments which comprise Γ are restricted to only q possible orientations, i.e. such a segment would lie along a line with equation

$$\mathbf{u}^{(m)}\cdot\mathbf{x} = b,$$

where $\mathbf{u}\in\mathbb{Z}[\lambda]^2$ is one of the vectors

$$\mathbf{u}^{(i)} = (\mathbf{C}^{\mathrm{T}})^i\cdot\begin{pmatrix}1\\0\end{pmatrix}, \qquad i = 0,\ldots,q-1,$$

where $\mathbf{C}^{\mathrm{T}}$ is the transpose of $\mathbf{C}$. We note that in the case of even q, we have

$$(\mathbf{C}^{\mathrm{T}})^{q/2} = -\mathbf{1},$$

hence the reflection symmetry

$$\mathbf{u}^{(i+q/2)} = -\mathbf{u}^{(i)},\ i = 0, \ldots, q/2-1.$$

In our investigation of the return-map hierarchy, the primary focus will be on piecewise isometric transformations of convex sets (of dimension zero, one or two) bounded by discontinuity lines and boundary lines of Ω. Such a convex set will be called a *domain*, and we will use the notation

$$\mathscr{D}(\mathbf{m},\mathbf{s},\mathbf{b}), \quad \mathbf{m} = (m_1, \ldots, m_n), \quad \mathbf{s} = (s_1, \ldots, s_n), \quad \mathbf{b} = (b_1, \ldots, b_n), \qquad (2.5)$$

to completely specify, in clockwise order, each boundary line in terms of an orientation index $m_i \in \{0, \ldots, q-1\}$ (corresponding to basis vector $\mathbf{u}^{(m_i)}$), a line coordinate $b_i \in \mathscr{M}(\tau)$, and a binary index $s_i = \pm 1$ distinguishing whether b_i is a lower or upper bound for $\mathbf{u}^{(m_i)} \cdot \mathbf{x}$ in the domain. For positive τ, the lower bounds are always inclusive, and the upper bounds exclusive, whereas for negative τ, the opposite is the case:

$$\begin{array}{llll} \tau > 0, s_i = -1: & \mathbf{u}^{(m_i)} \cdot \mathbf{x} \geq b_i, & \tau > 0, s_i = +1: & \mathbf{u}^{(m_i)} \cdot \mathbf{x} < b_i, \\ \tau < 0, s_i = -1: & \mathbf{u}^{(m_i)} \cdot \mathbf{x} > b_i, & \tau < 0, s_i = +1: & \mathbf{u}^{(m_i)} \cdot \mathbf{x} \leq b_i. \end{array}$$

Figure 2.8 illustrates these conventions in one of the renormalizable quadratic models.

The rectangular coordinates of a vertex of $\mathscr{D}(\mathbf{m},\mathbf{s},\mathbf{b})$ can be calculated, using elementary linear algebra, solving simultaneously the equations $\mathbf{u}^{(m_j)} \cdot \mathbf{x} = b_j$ and $\mathbf{u}^{(m_k)} \cdot \mathbf{x} = b_k$ for two boundary lines intersecting at the vertex. The coordinates are in $\mathbb{Q}(\lambda)$ but may not be elements of $\mathscr{M}(\tau)$. Because of the finite number of distinct orientations, however, they are in $\mathscr{M}(\tau)/a$, where a is the least common multiple of the determinants of the 2×2 matrices $(\mathbf{u}^{(m)}, \mathbf{u}^{(n)})$, $m, n \in \{0, 1, \ldots, q-1\}$.

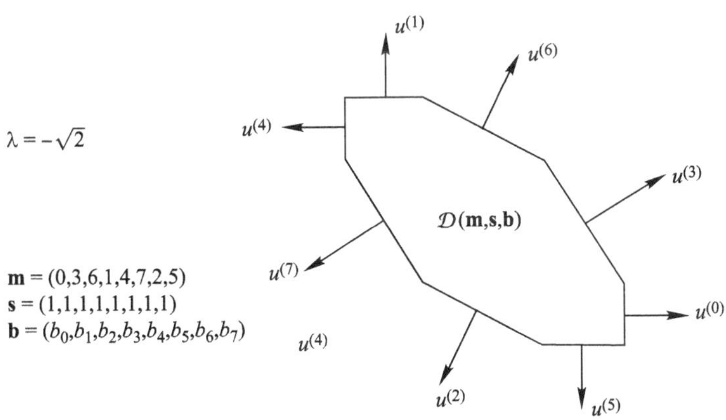

Fig. 2.8 Periodic domain in the model with $\lambda = -\sqrt{2}$, illustrating the notation $\mathscr{D}(\mathbf{m},\mathbf{s},\mathbf{b})$. Since $\tau = \sqrt{2}/2 > 0$ and $s_i = 1$ for all i, all of the bounds are exclusive upper bounds, and so the domain is an open octagon.

The area of a domain D with more than two vertices is easily calculated from the vertex coordinates. Suppose that $\mathbf{v}_0,\ldots,\mathbf{v}_{n-1}$ are the vertices of a domain, arranged in clockwise order. The area of the domain is then

$$\mathscr{A}(D) = \frac{1}{2}\sum_{i=1}^{n-2}(\mathbf{v}_{i+1}-\mathbf{v}_0)\times(\mathbf{v}_i-\mathbf{v}_0),$$

where the cross product $\mathbf{a}\times\mathbf{b}$ of a pair of 2-vectors is defined as the scalar quantity $a_x b_y - a_y b_x$.

2.2.3 Geometric transformations on domains

All of the dynamical mappings, scale transformations, and equivalence relations of domains relevant to our investigations belong to a group $\mathscr{G}$ of geometric transformations generated by a generalized rotation C (corresponding to application of the matrix $\mathbf{C}$), a generalized reflection R (provided that q is even), a scale transformation S_ω (where ω is a unique element in $\mathbb{Z}[\lambda]$), and a translations T_δ by an arbitrary 2-vector δ in $\mathbb{Z}[\lambda]^2$. The operations of these generating group elements on domains are the following:

$$C\mathscr{D}(\mathbf{m},\mathbf{s},\mathbf{b}) = \mathscr{D}(\mathbf{m}',\mathbf{s},\mathbf{b}), \qquad m'_k \equiv m_k - 1 \bmod q$$

$$R\mathscr{D}(\mathbf{m},\mathbf{s},\mathbf{b}) = \mathscr{D}(\tilde{\mathbf{m}},\mathbf{s},\mathbf{b}), \qquad \tilde{m}_k = \begin{cases} q/2-1-m_k, & 0\le m_k < q/2 \\ 3q/2-1-m_k, & q/2 \le m_k < q \end{cases}$$

$$S_\omega\mathscr{D}(\mathbf{m},\mathbf{s},\mathbf{b}_\infty+\mathbf{b}) = \mathscr{D}(\mathbf{m},\mathbf{s},\mathbf{b}_\infty+\omega\mathbf{b}), \qquad \omega \text{ a unit in } \mathbb{Z}[\lambda],\ \omega>0,\ \mathbf{b}_\infty\in\mathbb{Q}(\lambda)^2$$

$$T_\delta\mathscr{D}(\mathbf{m},\mathbf{s},\mathbf{b}) = \mathscr{D}(\mathbf{m},\mathbf{s},\mathbf{b}'), \qquad \delta\in\mathbb{Z}[\lambda]^2,\ b'_k = b_k + \mathbf{u}^{(k)}\cdot\delta.$$

We note that the scale transformations act on line coordinates relative to those of the scaling limit point. In the models under consideration, there are at most two such points. By choosing a frame of reference with the scaling limit at the origin, we can simplify the notation, eliminating the references to $\mathbf{b}_\infty$ in the scaling formulas.

The existence of inverses of the listed operations, needed to establish the group property of $\mathscr{G}$, is readily established. The fact that we only allow scaling by algebraic units is crucial here.

From the definitions, one easily sees that the generating elements of $\mathscr{G}$ have the following composition and commutation relations:

$$T_\delta\circ T_\varepsilon = T_{\delta+\varepsilon}, \qquad S_\omega\circ S'_\omega = S_{\omega\omega'}, \qquad R\circ R = 1,$$

$$S_\omega\circ C = C\circ S_\omega, \qquad S_\omega\circ R = R\circ S_\omega, \qquad R\circ C = C^{-1}\circ R,$$

$$G\circ T_{\mathbf{d}} = T_{G(\mathbf{d})}\circ G, \qquad G\in\{C,R,S_\omega\},$$

where, in the last relation, $G(\mathbf{d})$ refers to the pointwise action of G on the 2-vector $\mathbf{d}$, for example $C(\mathbf{d}) = \mathbf{C}\cdot\mathbf{d}$. Using the above formulas to reorder and combine factors, one can write any element $G \in \mathscr{G}$ in the canonical form

$$G(\kappa, \nu, \mathbf{d}, \omega) = S_\omega \circ T_\mathbf{d} \circ C^\nu \circ R^\kappa.$$

2.2.4 Scaling sequences

In Sect. 2.1 we introduced the basic ideas of return map dynamics and discussed informally the concept of scaling sequences of dressed domains in renormalizable models. We now provide a more precise definition and prove a theorem that will allow us to establish their central role in the quadratic models.

Definition 2.1. Let $\mathbf{x}_*$ be a point of Ω and ω a unit in $Z[\lambda]$ with magnitude < 1. A *scaling sequence* with limit point $\mathbf{x}_*$ and contraction factor ω is a sequence of dressed domains $\Delta(L) = (D(L), \rho(L)),\ L = 0, 1, \ldots\}$ such that $\mathbf{x}_* \in \overline{D(L)}$ and, if $\{(D_j(L), \rho_j(L)) : j = 0, \ldots, J\}$ is the partition of $\Delta(L)$ associated with return map $\rho(L)$, then, for all j,

$$D_j(L) = S_\omega D_j(L-1), \tag{2.6}$$
$$\rho_j(L) \circ S_\omega D_j(L-1) = S_\omega \circ \rho_j(L-1) D_j(L-1), \tag{2.7}$$
$$\rho_j(L) D_j(L) = \rho_{j_{\nu-1}}(L-1) \circ \cdots \circ \rho_{j_0}(L-1) D_j(L), \tag{2.8}$$

where ν and $j_0, \ldots, j_{\nu-1}$ are functions of j, independent of L,

$$\nu = \nu_j, \qquad j_t = p(j,t),\ t = 0, \ldots, \nu_j - 1. \tag{2.9}$$

The *path function* $p(j,t)$ associated with a scaling sequence specifies the level-independent return paths of the domains $D_j(L)$. Not surprisingly, it will be of prime importance when we introduce a more abstract distillation of the return map dynamics in the next chapter.

The spatial scaling of domains in the definition of a scaling sequence is inevitably tied up with temporal scaling as well, as we saw in the example of Sect. 2.1. Clearly the level-L return times $T_j(L)$ are related to those of level $L-1$ by

$$T_j(L) = \sum_{t=0}^{\nu_j - 1} T_{j_t}(L-1).$$

This can be reduced to a concise linear recursion relation by introducing the *incidence matrix* A_{ij}, which counts the number of times the return path of $D_j(L)$ passes through domain $D_i(L-1)$:

$$T_j(L) = \sum_i T_i(L-1) A_{ij}, \qquad A_{ij} = \sum_{t=0}^{\nu_j - 1} \delta_{i,p(j,t)},$$

where $\delta_{i,j}$ is the Kronecker delta. Asymptotically, the return time scales as α^L, $\alpha > 0$, where α is the largest eigenvalue of the transpose of the incidence matrix.

To exploit the concept of scaling sequences, we obviously need an efficient way of identifying them in specific models. This is provided by the following lemma (Kouptsov et al., 2002):

Lemma 2.1. *Let ω a unit in $Z[\lambda]$ with magnitude < 1 and let S_ω be the corresponding scaling operator centered at the point $\mathbf{x}_*$. Further, let D be a domain such that $\mathbf{x}_* \in \bar{D}$, and suppose that ρ is the first return map for D, partitioning it into subdomains D_j, $j = 0,\ldots,J-1$. The scaled domains $D(L) = S_{\omega^L}D$ and isometries $\rho(L) = S_{\omega^L} \circ \rho \circ S_{\omega^{-L}}$, for $L = 0,1,\ldots,$ form a scaling sequence with origin $\mathbf{x}_*$ and contraction factor ω, provided that $\rho(1)$ is a genuine return map for $D(1)$, and the respective partitions for levels 0 and 1 satisfy the renormalization equations* (2.6–2.8).

Proof. The proof is by induction. We assume that $\rho(L)$ and $\rho(L+1)$ are return maps for $D(L)$ and $D(L+1)$, respectively, such that the renormalization equations are satisfied for the level-L and $L+1$ partitions. We need to prove the same properties for levels $L+1$ and $L+2$.

We begin with the verification of (2.8) (the other two equations follow immediately from the definitions in the statement of the lemma). Thus, using the induction hypothesis,

$$\begin{aligned}
\rho_j(L+2)D_j(L+2) &= \rho_j(L+2) \circ S_\omega D_j(L+1) \\
&= S_\omega \circ \rho_j(L+1)D_j(L+1) \\
&= S_\omega \circ \rho_{j_{v-1}}(L) \circ \cdots \circ \rho_{j_0}(L)D_j(L+1) \qquad (2.10) \\
&= \rho_{j_{v-1}}(L+1) \circ \cdots \circ \rho_{j_0}(L+1) \circ S_\omega D_j(L+1) \\
&= \rho_{j_{v-1}}(L+1) \circ \cdots \circ \rho_{j_0}(L+1)D_j(L+2).
\end{aligned}$$

Note that the third equality is obtained by commuting S_ω through the chain of partial return maps $\rho_j(L)$. This is not simply a matter of expressing $\rho_j(L)$ in the form of an isometry $T_\delta C^v$ and applying the group commutation relations to obtain

$$S_\omega \circ T_{\mathbf{d}} \circ C^v = T_{\omega\mathbf{d}} \circ C^v S_\omega.$$

One must also check, as we have done, that the range of each mapping in the chain is contained in the domain of the mapping to the left.

To complete the proof, we need to show that $\rho(L+2)$ is in fact a first return map for $D(L+2)$. From (2.10) and the induction hypothesis, $\rho_j(L+2)$ decomposes into v_j linked orbits of the elementary map K, with a total return time of

$$T_j(L+2) = \sum_{k=0}^{v_k-1} T_k(L+1).$$

We must show that each of the $T_j(L+2)$ domains on the orbit, with the exception of the initial and final one, is disjoint from $D(L+2)$. From the induction hypothesis all of the relevant domains, with the exception of

$$E_k(L+2) = \rho_{j_k}(L+1) \circ \cdots \circ \rho_{j_0}(L+1) D_j(L+2), \quad 0 \le k < \nu_j - 1,$$

are disjoint from $D(L+1)$, hence from $D(L+2) \subset D(L+1)$. The $E_k(L+2)$ are in $D(L+1)$, but

$$E_k(L+2) \cap D(L+2) = S_\omega E_k(L+1) \cap D(L+1) = \emptyset.$$

□

2.2.5 Periodic orbits

Suppose that Π is a periodic domain contained in the lowest-level member $D(0)$ of a scaling sequence. This means that $K^\tau \Pi = \Pi$ for some (nonzero, minimal) return time τ. Without loss of generality, we assume $\Pi \subset D_{j_0}(0)$ for some $j_0 \in \{0, \ldots, J-1\}$, so that there exists a level-0 return orbit

$$\Pi = \rho_{j_{\nu-1}} \circ \cdots \circ \rho_{j_0} \Pi$$

for some finite sequence of indices $j_k \in \{0, \ldots, J-1\}$, $k = 0, \ldots, \nu - 1$. Clearly Π can be promoted to an infinite sequence of periodic domains $\Pi(L)$ by multiple application of the scaling operator S_ω:

$$\Pi(L) = S_\omega^L \Pi, \quad L = 0, 1, \ldots,$$

with level-L return orbit

$$\Pi(L) = \rho_{j_{\nu-1}}(L) \circ \cdots \circ \rho_{j_0}(L) \Pi(L)$$

and return time

$$\tau(L) = \sum_{k=0}^{\nu-1} T_{j_k}(L).$$

Here ν and the return path $j_0, \ldots, j_{\nu-1}$ depend on Π, but not on the level L.

We note that there is a distinction between the *return time* of a periodic domain and the *minimal period* of its points under K. If a periodic domain Π has rotational symmetry (i.e. $C^\mu \Pi = \Pi$ for $\mu < q$), the centroid (mean position of the vertices) will have period τ, while the period of all other points in Π will be the least common multiple of τ and q.

2.2.6 Recursive tiling

As pointed out earlier in this chapter, the existence of a scaling sequence is essential to what we mean by renormalizability of a piecewise isometric system, but it may not be the whole story. There must be some sense in which the return map dynamics gives a *complete* description of the dynamics on Ω, at least on the complement of the discontinuity set. We now turn to a precise definition of this concept, which we call the *recursive tiling* property.

A piecewise isometric mapping of the square Ω is said to satisfy the recursive tiling property if there exist finitely many scaling sequences

$$\{(D^{(\sigma)}(L^{(\sigma)}),\rho^{(\sigma)}(L^{(\sigma)})) : L^{(\sigma)} = 0,1,\ldots\},\ \sigma = 0,\ldots,N-1,$$

such that, for any choice of levels $L^{(\sigma)}$, the return orbits of the subdomains $D_j^{(\sigma)}(L^{(\sigma)})$, together with those of the periodic domains $\Pi_k^{(\sigma)}(M^{(\sigma)})$ of level $M^{(\sigma)} < L^{(\sigma)}$ and a finite number of exceptional periodic domains, completely tile Ω, apart from the measure-zero set of domain boundaries. Since orbits on the discontinuity set can be ignored, the task of establishing the recursive tiling property reduces to summing up the areas of the tiles comprising the return orbits to check that the total is equal to τ^2, the area of Ω.

In the present chapter we will emphasize that a stronger definition of recursive tiling is within reach, provided that one is willing to expend the (considerable) additional effort to do a proper accounting of the return orbits on the discontinuity set. Two quadratic models, where this has been achieved (Kouptsov et al., 2002), will be studied in detail below. The following definitions and lemma state a concise set of tests for the strong form of recursive tiling.

Definition 2.2. Given a line segment γ in the plane, the skew length of γ, written $\mathscr{L}(\gamma)$, is the $\mathscr{Q}$-metric length of the segment.

Definition 2.3. The weighted perimeter function $\mathscr{W}(D)$ assigns to a domain D a number in $\mathbb{Q}(\lambda)$ as follows:

1. If D is an isolated point, $\mathscr{W}(D) = 0$.
2. If D is a line segment , $\mathscr{W}(D)$ is its skew length, $\mathscr{L}(D)$.
3. If D is a nondegenerate polygon, $\mathscr{W}(D)$ is a weighted sum over the skew lengths of its edges, with weighting factor $1/2$ if the edge is included in D, and $-1/2$ if it is excluded.

We note that with the representation of a polygonal domain as $\mathscr{D}(\mathbf{m},\mathbf{s},\mathbf{b})$, the inclusion or exclusion of an edge can be read off from the sign of the relevant component s_i of $\mathbf{s}$.

Lemma 2.2. *To prove the recursive tiling property (strong form) of a quadratic model equipped with a scaling sequence* $(D(L),\rho(L))$, $L = 0,1,\ldots$ *with a finite number of associated sequences of periodic domains* $\Pi_j(L)$*, and a finite number of non-scaling periodic domains* $\Pi_{j0}(\Omega)$ *whose orbits are disjoint from* $D(0)$*, it is sufficient to verify the following sum rules*:

1. *Area sum, tiling of* Ω.

$$\sum_{j=0}^{J-1} \nu_j(0)\mathscr{A}(D_j(0)) + \sum_{j_0=0}^{J_0-1} \tilde{\nu}_{j_0}(\Omega)\mathscr{A}(\Pi_{j_0}(\Omega)) = \mathscr{A}(\Omega).$$

2. *Area sum, tiling of $D(0)$.*

$$\sum_{j=0}^{J-1} \nu_j(1)\mathscr{A}(D_j(1)) + \sum_{\tilde{j}=0}^{\tilde{J}} \tilde{\nu}_{\tilde{j}}(0)\mathscr{A}(\Pi_{\tilde{j}}(0)) = \mathscr{A}(D(0)).$$

3. *Weighted perimeter sum, tiling of Ω.*

$$\sum_{j=0}^{J-1} \nu_j(0)\mathscr{W}(D_j(0)) + \sum_{j_0=0}^{J_0-1} \tilde{\nu}_{j_0}(\Omega)\mathscr{W}(\Pi_{j_0}(\Omega)) = \mathscr{W}(\Omega).$$

4. *Weighted perimeter sum, tiling of $D(0)$.*

$$\sum_{j=0}^{J-1} \nu_j(1)\mathscr{W}(D_j(1)) + \sum_{\tilde{j}=0}^{\tilde{J}} \tilde{\nu}_{\tilde{j}}(0)\mathscr{W}(\Pi_{\tilde{j}}(0)) = \mathscr{W}(D(0)).$$

5. *Vertex coverage, tiling of Ω. The orbits of the $D_j(0)$ and $\Pi_{\tilde{j}}(\Omega)$ must cover all vertices not lying on the excluded boundaries of Ω.*
6. *Vertex coverage, tiling of $D(0)$. The orbits of the $D_j(1)$ and $\Pi_{\tilde{j}}(0)$ must cover all vertices not lying on the excluded boundaries of $D(0)$.*

For the weak form of recursive tiling, the area sums suffice.

Proof. Thanks to the scaling of areas and lengths, it will be sufficient to prove (a) the tiling of Ω by the return orbits of $D_j(0)$, $j = 0,\dots,J$ and of the periodic domains which do not intersect $D(0)$, and (b) the tiling of $D(0)$ by the return orbits of $D_j(1)$, $j = 0,\dots,J$ and of the periodic domains of $D(0)$ which do not intersect $D(1)$. Since the union of relevant return orbits in Ω comprises a set of mutually disjoint convex polygons, the area sum rules imply coverage up to a set restricted to the domain boundaries, not including those which lie along the excluded boundaries of Ω, i.e. along $x = \tau$ or $y = \tau$. The weighted perimeter sums, on the other hand, guarantee full coverage of the domain boundary segments, apart from a finite number of isolated points. Having complete coverage of such vertices completes the proof of 100% coverage. Analogous arguments establish the complete coverage of $D(0)$. □

For models with more than one scaling sequence (such as Model II below), Lemma 2.2 can immediately be generalized. The modifications to the statement and proof are obvious, and we omit the details. Another rather simple, but useful, generalization (needed for Model III below) involves the technique of *telescoping*. This refers to the situation where we have a sequence of dressed domains $\Delta(L)$, $L = 0,1,2,\dots$, which possess the recursive tiling property but where the renormalization equations relate non-consecutive levels. For example, for some fixed k, levels L and $L+k$, $L=0,1,2,\ldots$, may be so linked. In such a case, we might have k inequivalent scaling sequences, all governed by the same scale factor ω. Provided that the full sequence satisfies the area and perimeter sum rules of Lemma 2.2, not just for levels 0 and 1, but for levels $0,\ldots,k$, it is clear that the hierarchy can be collapsed ("telescoped") into a true scaling sequence

$$\hat{\Delta}(N) = \Delta(Nk),\ N = 0,1,2,\ldots.$$

This sequence will satisfy all the hypotheses of Lemma 2.2, and hence will be recursively tiled with respect to the telescoped sequence of levels. Of course, the return paths may become extraordinarily large for the telescoped hierarchy, which is why one might prefer to work with the original one with its leap-frogging scaling relations. This will be our preference in the formulation of Model III below.

The union of all return orbits initiated in the base domain $D(0)$ of a scaling sequence is an invariant set containing, in the models of interest, both periodic and aperiodic points. The set of aperiodic points on discontinuity-avoiding orbits is obviously contained in every set of the form

$$\mathscr{C}(L) = \bigcup_{j=0}^{J-1} \bigcup_{t=0}^{\nu_j-1} \rho(L-1)^t D_j(L),$$

and hence the area occupied by these points is bounded above by the area $\mathscr{A}(\mathscr{C}(L))$ for all L. But

$$\mathscr{A}(\mathscr{C}(L)) \sim (\omega^2\alpha)^L \text{ for } L \to \infty,$$

where α is the largest eigenvalue of the incidence matrix. In all of the quadratic models $\omega^2\alpha < 1$, and so the aperiodic points, including those on the discontinuity set, occupy a set of measure zero. Consequently, the periodic domains within a scaling sequence have full measure. The same property extends to the full set of periodic domains within Ω: they tile the square apart from a set of measure zero.

In what follows, we will refer to the zero-measure complement of the periodic domains as the *exceptional set*, and to the subset of discontinuity-avoiding aperiodic points as the *residual set*. Due to the decrease without limit of the constituent domains of the covering sets $\mathscr{C}(L)$ as L increases, we conclude that each point in the residual set is arbitrarily close to a point of the discontinuity set Γ, and hence the exceptional set can be identified with the closure of Γ.

2.2.7 Computer-assisted proofs

In Kouptsov et al. (2002), nine quadratic models were shown to be renormalizable in the sense discussed above. Eight of these comprise the family of piecewise isometric models described at the beginning of Sect. 2.2, motivated by the role of these maps in models of Hamiltinian round-off (Lowenstein et al., 1997; Lowenstein and

Vivaldi, 2000). The ninth, which differs from the rest in having the fixed point at the center of Ω rather than at the corner $(0,0)$, was the subject of an important article by Adler et al. (2001) on two-dimensional piecewise isometries.

Computer assisted proofs were given by Kouptsov et al. (2002) to establish the existence of a complete set (in the sense of recursive tiling) of scaling sequences in all nine cases. A summary of the most important parameters is given in Table 2.1. Note that our parameter λ, which appears in the generalized rotation matrix **C**, is the algebraic conjugate of the parameter λ of Kouptsov et al. (2002). This reflects the shift of emphasis from round-off models to piecewise isometries and kicked oscillator maps. The last column of the table lists the asymptotic temporal scaling factor α, the largest eigenvalue of the transpose of the incidence matrix A.

Table 2.1 Scaling parameters for the renormalizable quadratic models of Kouptsov et al. (2002).

λ	p/q	τ	Ω	ω	$\mathbf{x}_*$	α
$(-1+\sqrt{5})/2$	$1/5$	$(1-\sqrt{5})/2$	$(\tau,0]^2$	$1+\tau$	$(0,0)$	4
$(-1-\sqrt{5})/2$	$2/5$	$(1+\sqrt{5})/2$	$[0,\tau)^2$	$2-\tau$	$(0,0)$	4
$(1+\sqrt{5})/2$	$1/10$	$(-1-\sqrt{5})/2$	$(\tau,0]^2$	$2+\tau$	$(0,\tau)$	4
$(1-\sqrt{5})/2$	$3/10$	$(-1+\sqrt{5})/2$	$[0,\tau)^2$	$1-\tau$	(τ,τ)	4
$-\sqrt{2}$	$1/8$	$\sqrt{2}/2$	$[0,\tau)^2$	$3-4\tau$	$(0,\tau)$	9
$\sqrt{2}$	$3/8$	$-\sqrt{2}/2$	$(\tau,0]^2$	$3+4\tau$	(τ,τ)	9
$\sqrt{3}$	$1/12$	$-\sqrt{3}/3$	$(\tau,0]^2$	$2+3\tau$	(τ,τ)	4
$\sqrt{3}$	$1/12$	$-\sqrt{3}/3$	$(\tau,0]^2$	$2+3\tau$	$(41+72\tau,41+72\tau)$	4
$-\sqrt{3}$	$5/12$	$\sqrt{3}/3$	$[0,\tau)^2$	$7-12\tau$	(τ,τ)	25
$\sqrt{2}$	$3/8$	$-\sqrt{2}$	$(\tau,-\tau]^2$	$3+4\tau$	$(-1-\tau,\tau)$	9

Before getting into a detailed review of some of these results, it is important to make clear the role which computer hardware and software played in obtaining them. This is particularly important in light of our claim that the principal statements regarding renormalization deserve to be treated as possessing a level of mathematical rigor comparable to that of other propositions proved without the assistance of electronic devices. Computer calculations enter into our investigations at three stages as follows

Phase 1: exploration

Each of the piecewise isometric models which we deal with is a highly complex mathematical environment which, in the first stage, we explore using the least sophisticated tools: direct iteration of mappings using machine-precision floating-point representations of the real numbers, motivated and assisted by computer graphics. The existence (or not) of self-similarity (i.e. apparent fractal structure in the phase portrait), as well as the likely locations of scaling fixed points, are usually, but not always apparent at this stage.

Phase 2: derivation

This is essentially the heuristic approach described informally at the beginning of this chapter, carried out, however, without making numerical approximations. This entails treating τ as an abstract symbol, with polynomials of degree greater than two reduced to the form $r+s\tau,\ r,s\in\mathbb{Q}$ via application of the quadratic equation satisfied by τ. In order to efficiently carry out the construction of fairly extensive branches of the various return map trees, specialized software in the C programming language was developed. This included a library of functions to handle the arithmetic of arbitrarily large integers, as well as of algebraic numbers in $\mathbb{Z}[\lambda]$ and $\mathbb{Q}(\lambda)$. Other functions were invented to handle the mapping and intersection of domains. Finally, programs were developed to construct recursively the return map partitions of any level L.

Phase 3: computer assisted proof

While the constructions of Phase 2 reliably produce a detailed accounting of the scaling sequences and periodic domains of the nine models listed in Table 2.1, they do not constitute a *computer assisted proof*, even if they are a meticulously executed *computational experiment*. The difference lies in the relative transparency of the arithmetic and algebraic procedures. For example, as part of a non-computational proof, a mathematician would find acceptable carrying out "by hand" any number of elementary operations involving the addition, subtraction, multiplication and division of rational numbers and polynomials with rational coefficients. Presumably he or she would have no objections to having a commercially available computer running (say) Mathematica carry out the same set of procedures as part of a proof. Should the same level of trust extend to a custom-designed function used to calculate the intersection of two domains, yielding reliable results even in cases where the intersection is a line segment or a single point with coordinates in $\mathbb{Q}(\lambda)$? Presumably not.

Fortunately, it is not necessary for the proof of a mathematical theorem to be constructive. We need only be sufficiently careful in using the computer to verify the multitude of results obtained by the less than rigorous methods of Phase 2. We do this using a small number of very short procedures, each of which is sufficiently simple and transparent that it could easily be performed "by hand". The role of computer is to perform thousands of these elementary operations in succession within a fraction of a second (rather than the much more extended time period, sometimes exceeding the age of the universe, needed for a human being to do the same). What makes this simplicity possible is the fact that once return paths and the subdomains of a return map partition have been specified in detail, to prove the correctness of those return paths and partition involves only mapping and checking inclusion relations, but no intersections of domains and no branching of orbits.

For the proof of the existence and derived properties of the listed scaling sequences and their associated periodic domains, the following elementary procedures (all but two of which are made explicit as Mathematica functions in an appendix of Kouptsov et al. (2002)) are employed:

`signum`$(a+b\tau)$ Returns 0 if $a=b=0$, 1 if $a+b\tau>0$, and -1 if $a+b\tau<0$.

`vertex`(m_1,m_2,b_1,b_2) Calculates the intersection point of the lines $u^{(m_j)}\cdot\mathbf{x}=b_j$, $j=1,2$.

`vertexconstraint`(v,m,s,b) Returns $-1,1,$or 0, depending on whether the vertex v lies inside, outside, or on the boundary of the half-plane $su^{(m)}\cdot\mathbf{x}<sb$.

`vertices`(D) Returns the set of vertices of the domain D, in cyclic order.

`disjoint`(D_1,D_2) Returns True if domains D_1 and D_2 are disjoint.

`subdomain`(D,D_1) Returns True if domain D is a subdomain of domain D', False otherwise.

`equal`(D,D') Returns True if domain D is equal to domain D', False otherwise.

`area`(D) Returns the (Euclidean metric) area of domain D.

`length`$(\mathbf{x},\mathbf{y})$ Returns the ($\mathscr{Q}$-metric) length of the line segment connecting $\mathbf{x}$ to $\mathbf{y}$.

`perimeter`(D) Returns the ($\mathscr{Q}$-metric) perimeter of the domain D.

`rescale`(a,D) Returns a domain equal to domain D with all bounds rescaled by a factor $a\in\mathbb{Z}[\lambda]$.

`rotate`(ν,D) Returns $C^{\nu}D$.

`translate`$(\mathbf{d},D)$ Returns $T_{\mathbf{d}}D$.

`map`$(\nu,\mathbf{d},D)$ Returns $T_{\mathbf{d}}\circ C^{\nu}D$.

`phimap`(D) Implements the map K on the domain D.

`rho`(L,j,D) Implements the map $\rho_j(L)$ on the domain D.

`returntest1`(L,j) Tests whether $\rho_j(L)$, as defined in terms of iterations of K, is in fact the return map for $D_j(L)$. Returns True or False.

`returntest2`(L,j) Tests whether $\rho_j(L)$, defined in terms of iterations of $\rho(L-1)$, is in fact the return map for $D_j(L)$. Returns True or False.

`returntest3`(L,j) Tests whether $\Pi_j(L)$ is a periodic domain with a specified return path. Returns True or False.

The only one of these functions which is not straightforward to program is `signum`$(a+b\tau)$ which we must implement without resorting to a floating-point approximation of the quadratic irrational τ. The trick is to use the continued-fraction expansion (Khinchin, 1964) of τ, whose even and odd convergent satisfy the recursion relations, for $k>0$,

$$\frac{n_{2k}}{d_{2k}}=\frac{f_0n_{2k-1}+n_{2k-2}}{f_0d_{2k-1}+d_{2k-2}},\qquad \frac{n_{2k+1}}{d_{2k+1}}=\frac{f_1n_{2k}+n_{2k-1}}{f_1d_{2k}+d_{2k-1}},$$

with recursion coefficients and initial approximants given in Table 2.2:

Table 2.2 Recursion coefficients and initial approximants.

p/q	f_0	f_1	n_0	d_0	n_1	d_1
1/5	1	1	−1	1	−1	2
2/5	1	1	1	1	2	1
1/8	2	2	−1	1	−2	3
3/8	2	2	1	1	2	3
1/10	1	1	−1	1	−2	1
3/10	1	1	1	1	1	2
1/12	2	2	−1	1	−1	2
5/12	2	2	1	1	1	2

The rational approximants $m_{2k} = a + bn_{2k}/d_{2k}$ and $m_{2k+1} = a + bn_{2k+1}/d_{2k+1}$ bracket $a + b\tau$, and so to calculate $\mathtt{signum}(a + b\tau)$ for $b \neq 0$, it suffices to keep iterating the recursion relation until m_{2k} and m_{2k+1} are found to have the same sign. The latter then gives the desired sign of $a + b\tau$.

The two functions added to the original toolbox of the "computer assistant" of Kouptsov et al. (2002) are `length` and `perimeter`. These will be needed below in order to establish the strong form of recursive tiling in two of the models introduced below. In these procedures, the measurement of the $\mathcal{Q}$-distance between two end-points of a line segment is measured by first rotating the segment by application of the matrix $\mathbf{C}$ until it is horizontal, then taking the difference of the x components. This works, since $\mathbf{C}$-rotation leaves the $\mathcal{Q}$-distance unchanged, and on a horizontal line Euclidean distance coincides with $\mathcal{Q}$-distance.

2.3 Three quadratic models

We now examine in detail three piecewise isometric mappings of the square with quadratic irrational parameters. The first of these, Model I, is fairly representative of the quadratic models with rotation number p/q, $q < 12$. The second, Model II, exhibits the highest level of complexity and the least transparent renormalization structure of any of the quadratic examples studied, and thus is an excellent guard against too facile generalizations.

The third model, Model III, originally introduced by Adler et al. (2001) and further developed by Kouptsov et al. (2002), is unique among our examples in its symmetrical placement of the rotation center, and provides a beautiful example of rich dynamics on the discontinuity set. Apart from their intrinsic interest as piece-wise isometries of the square, all three of the selected models are equally fascinating when lifted to the infinite plane as kicked-oscillator maps. We will leave this part of the story until later chapters.

In the following pages we present in a concise format the core data of the three models, with accompanying figures to help the reader gain an intuitive appreciation of the models without having to wade through the details. The full data sets needed for computer assisted proofs of renormalizability, as well as the kicked-oscillator applications of later chapters, have been placed in Appendix A. These data are in one sense more elaborate than one might think necessary: the partition of Ω into subdomains (the so-called *generating partition*) is carried out in such a way that all two-dimensional domains are open, i.e. do not contain their any of their edges. Similarly, the one-dimensional domains do not contain their endpoints. This greatly facilitates the study of the discontinuity-set dynamics and is well worth the extra bookkeeping.

2.3.1 *Model I*

Parameter summary

λ	p/q	τ	Ω	ω	$\mathbf{x}_\infty$	α
$(1+\sqrt{5})/2$	$1/10$	$(-1-\sqrt{5})/2$	$(\tau,0]^2$	$2+\tau$	$(0,\tau)$	4

Orientation co-vectors

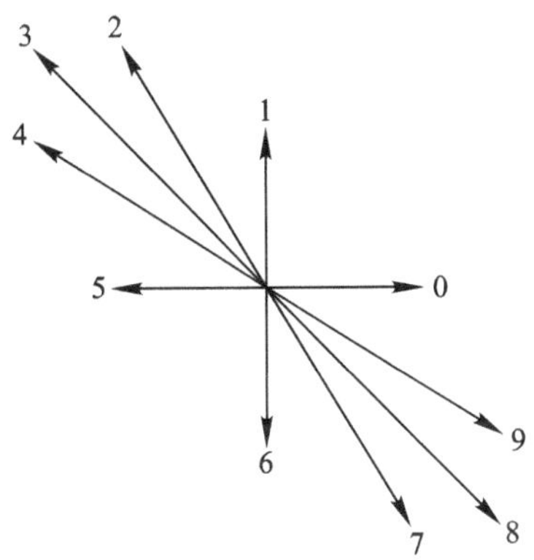

m	$\mathbf{u}^{(m)} = -\mathbf{u}^{(m+5)}$
0	$(1, 0)$
1	$(0, 1)$
2	$(-1, -\tau)$
3	$(\tau, -\tau)$
4	$(\tau, 1)$

Map K on the generating domain Ω

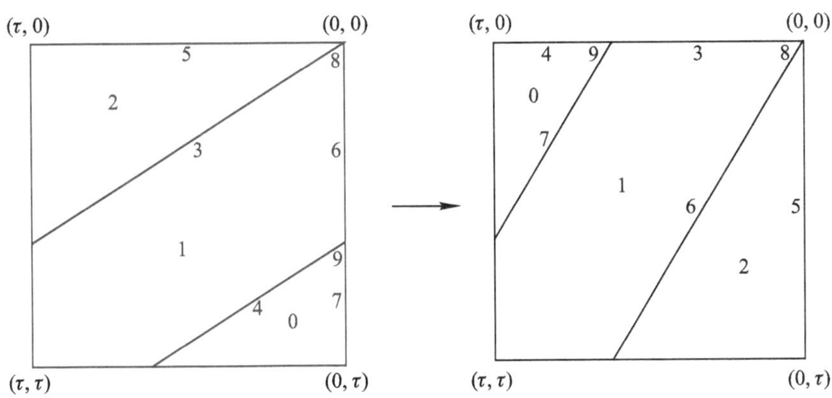

Level L atoms, return paths

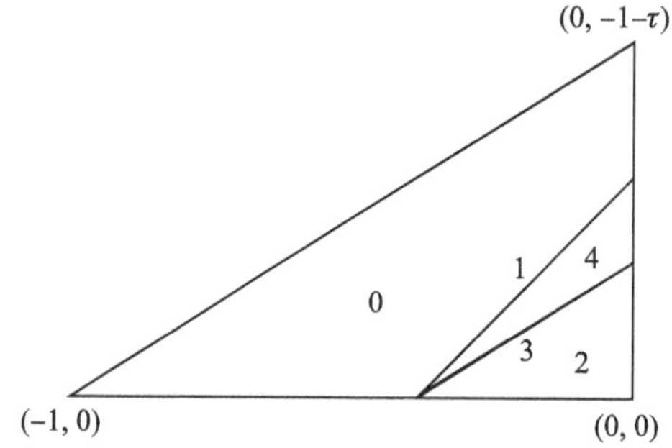

$D_j(L)$ return paths:

$p(0) = (2, 0)$

$p(1) = (2, 0, 3, 0)$

$p(2) = (2, 0, 4, 0)$

$p(3) = (2, 0, 4, 0, 3, 0, 4, 0)$

$p(4) = (2, 0, 4, 0, 4, 0, 4, 0)$

Level L periodic domains, return paths

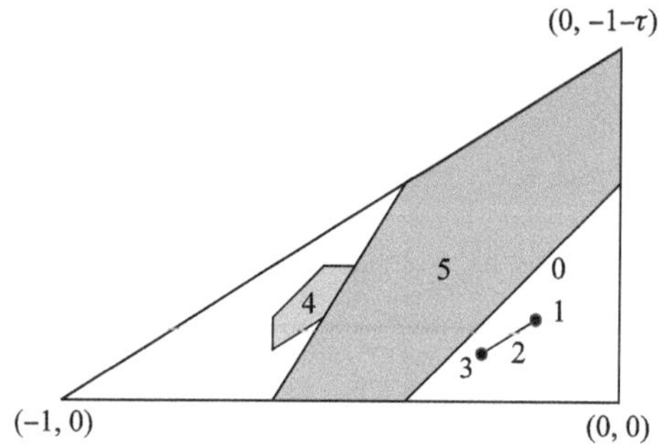

$\Pi_j(L)$ return paths:

$p(0) = (1, 0)$

$p(1) = (3, 0)$

$p(2) = (3, 0, 4, 0, 4, 0, 4, 0)$

$p(3) = (3, 0, 4, 0)$

$p(4) = (0, 4)$

$p(5) = (0)$

Incidence matrix

$$\mathbf{A} = \begin{pmatrix} 1 & 2 & 2 & 4 & 4 \\ 0 & 0 & 0 & 0 & 0 \\ 1 & 1 & 1 & 1 & 1 \\ 0 & 1 & 0 & 1 & 0 \\ 0 & 0 & 1 & 2 & 3 \end{pmatrix}$$

Recursive tiling of Ω by atoms of $D(0)$

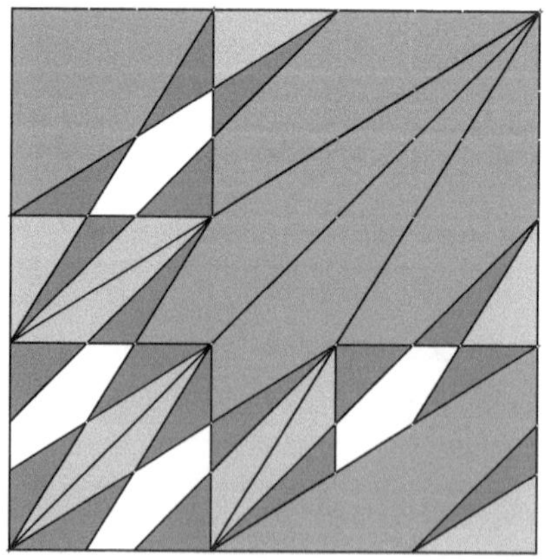

Recursive tiling of $D(0)$ by atoms of $D(1)$

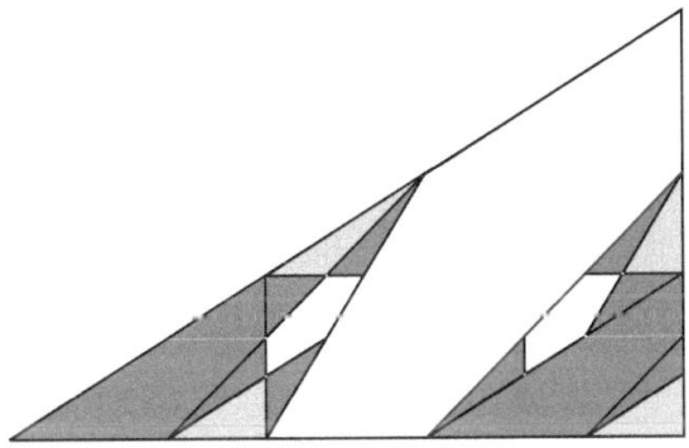

2.3.2 *Model II*

Parameter summary

λ	p/q	τ	Ω	ω	$\mathbf{x}_\infty$	α
$\sqrt{3}$	$1/12$	$-\sqrt{3}/3$	$(\tau,0]^2$	$2+3\tau$	(τ,τ) $(41+72\tau,41+72\tau)$	4 5

Orientation co-vectors

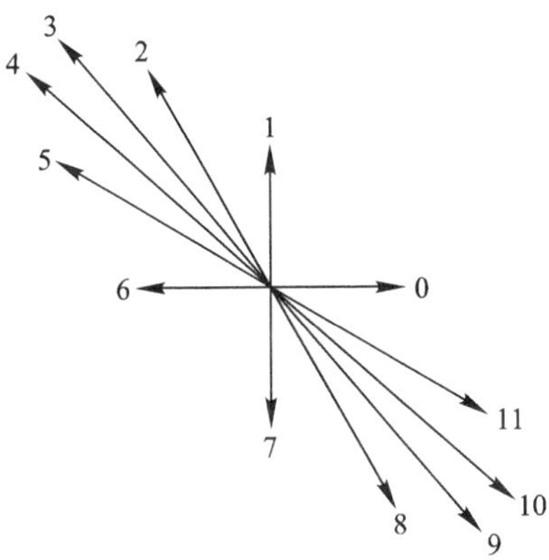

m	$\mathbf{u}^{(m)} = -\mathbf{u}^{(m+6)}$
0	$(1, 0)$
1	$(0, 1)$
2	$(-1, -3\tau)$
3	$(3\tau, 2)$
4	$(-2, -3\tau)$
5	$(3\tau, 1)$

Map K on the generating domain Ω

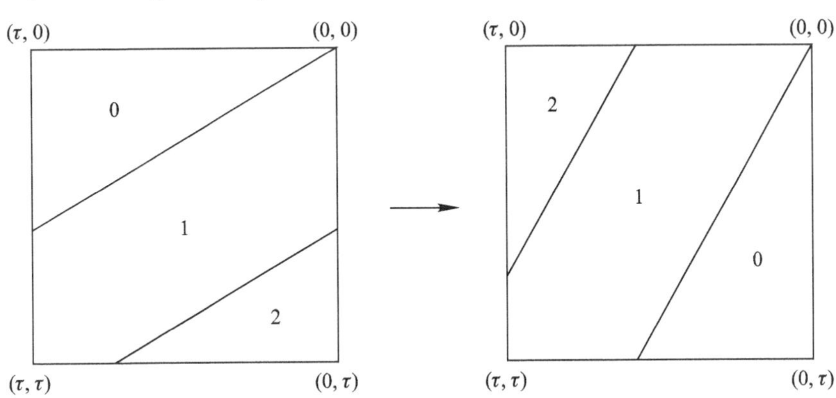

Return map partitions of $D^{(\mathbf{A})}(0)$ and $D^{(\mathbf{B})}(0)$

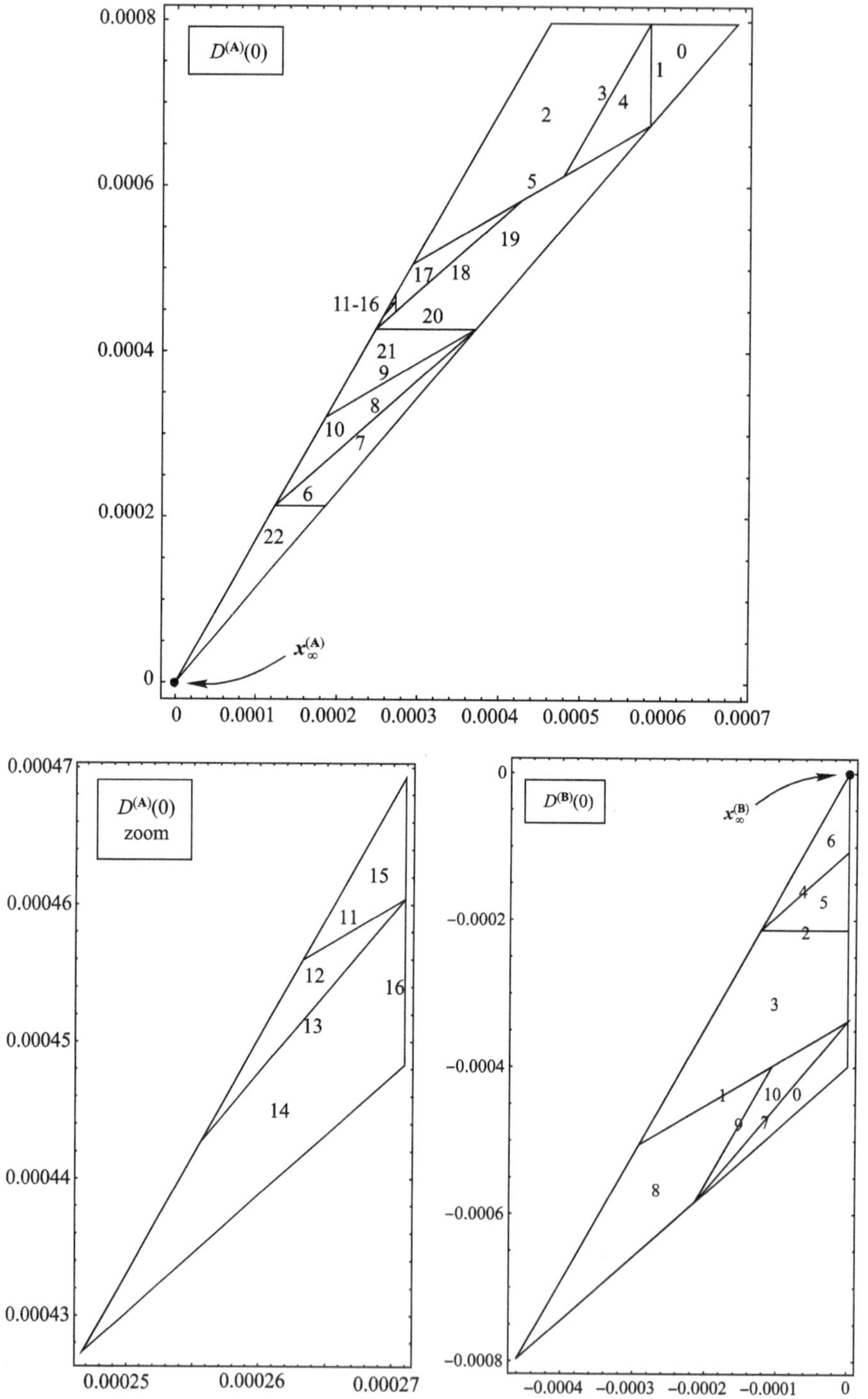

Tiling of $D^{(\mathbf{A})}(0)$.[1]

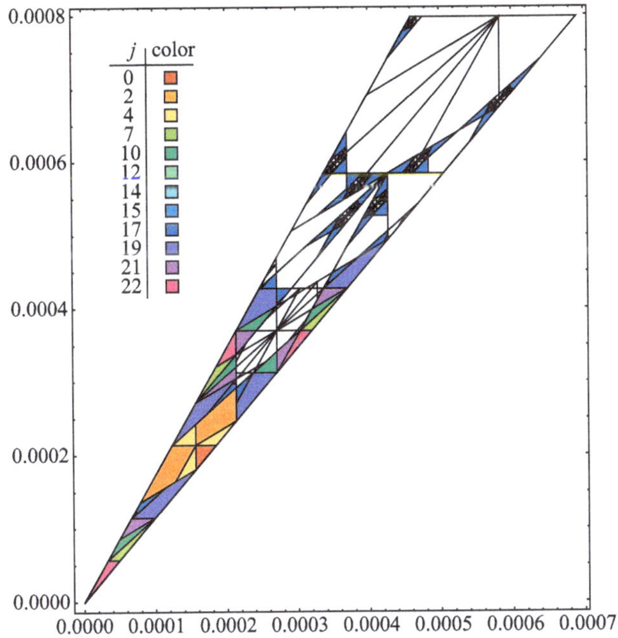

Tiling of $D^{(\mathbf{B})}(0)$.[1]

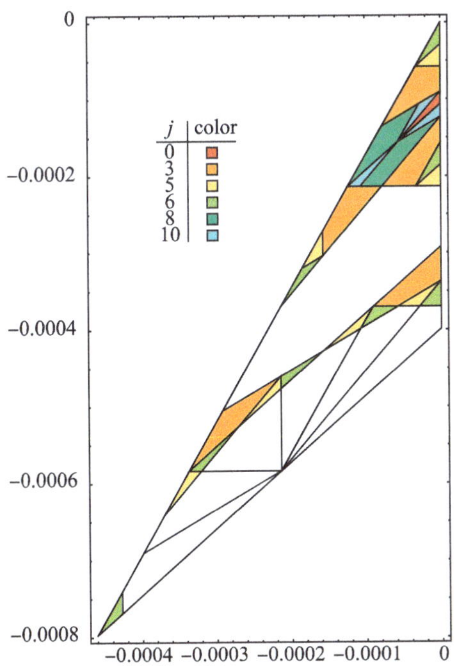

[1] See color figures at the end of the book.

2.3.3 *Model III*

Parameter summary

λ	p/q	τ	Ω	ω	$\mathbf{x}_\infty$	α
$\sqrt{2}$	$1/8$	$-\sqrt{2}/2$	$(\tau, -\tau]^2$	$3+4\tau$	$(-1-\tau, \tau)$	9

Orientation co-vectors

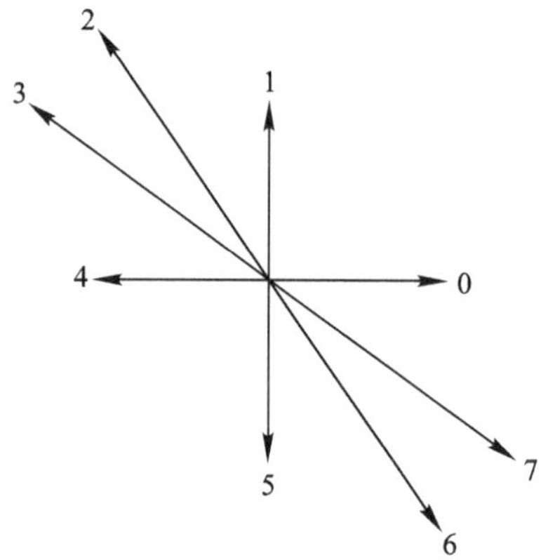

m	$\mathbf{u}^{(m)} = -\mathbf{u}^{(m+4)}$
0	(1, 0)
1	(0, 1)
2	(-1, -2τ)
3	(2τ, 1)

Map K on the generating domain Ω

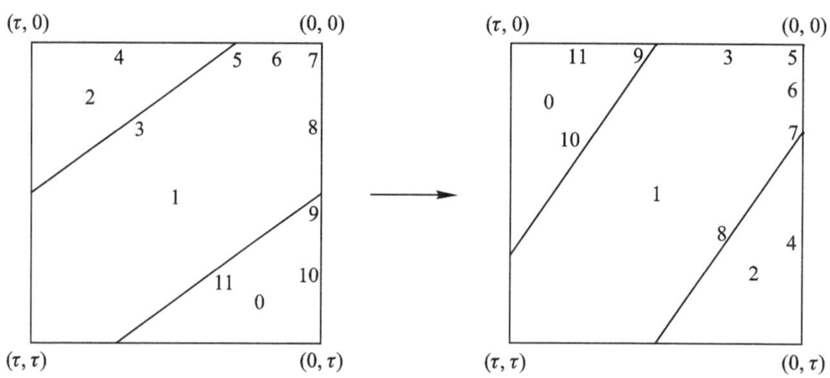

Prescaling domain

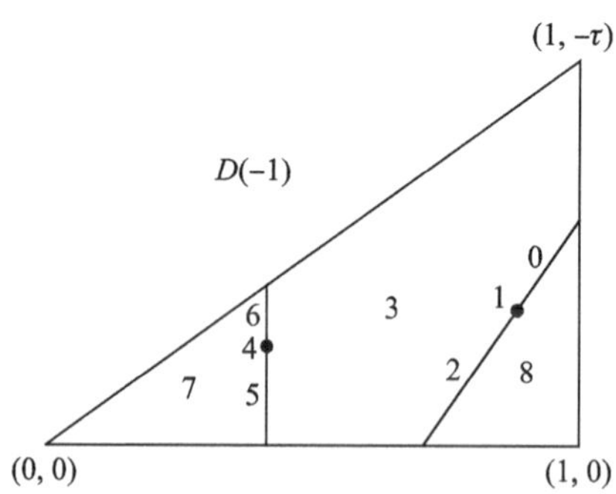

Level L atoms

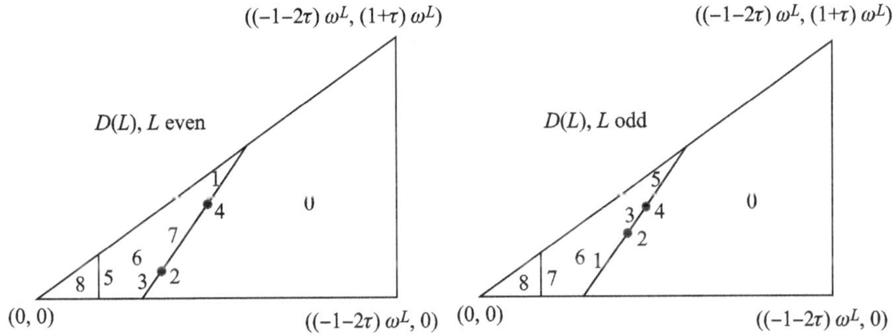

Incidence matrix

$$\mathbf{A} = \begin{pmatrix} 1 & 13 & 35 & 39 & 50 & 49 & 12 & 23 & 24 \\ 0 & 1 & 3 & 2 & 3 & 3 & 0 & 1 & 0 \\ 0 & 0 & 1 & 0 & 0 & 0 & 0 & 0 & 0 \\ 0 & 0 & 1 & 1 & 1 & 1 & 0 & 1 & 0 \\ 0 & 0 & 0 & 0 & 1 & 0 & 0 & 0 & 0 \\ 0 & 2 & 3 & 2 & 3 & 2 & 0 & 1 & 0 \\ 0 & 2 & 6 & 8 & 10 & 12 & 5 & 6 & 8 \\ 0 & 1 & 2 & 1 & 1 & 0 & 0 & 1 & 0 \\ 1 & 1 & 1 & 1 & 1 & 1 & 1 & 1 & 1 \end{pmatrix}$$

Level L periodic domains

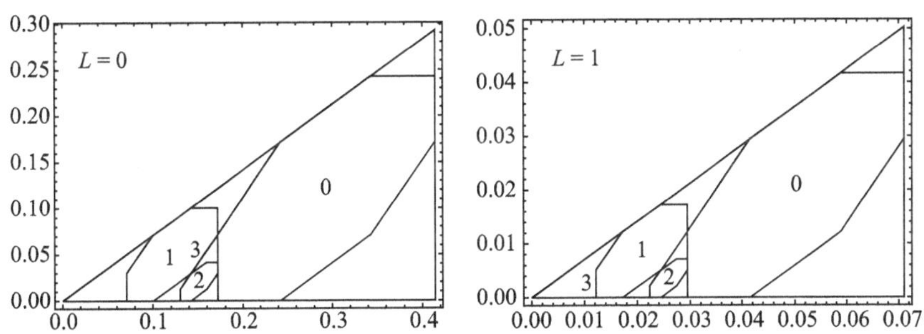

Recursive tiling

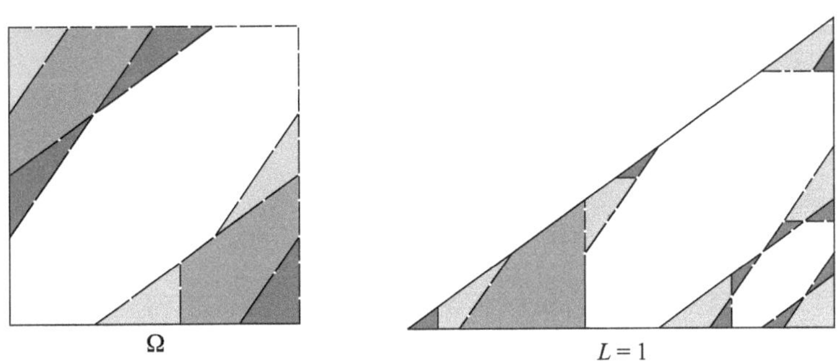

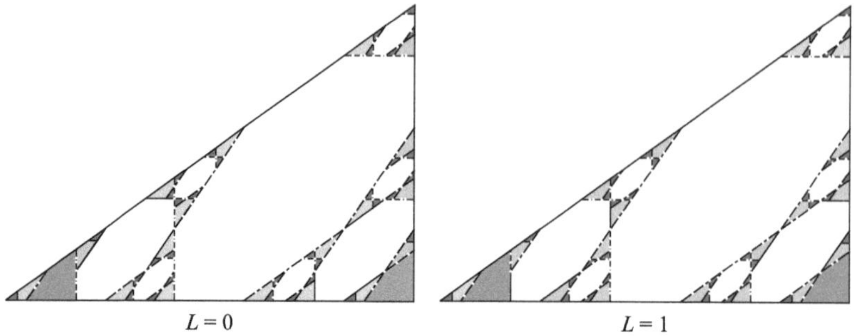

$L=0$ $L=1$

2.4 Proof of renormalizability

The proof of the renormalizability of Models I, II, and III consists of verifying, with computer assistance, the hypotheses of Lemmas 2.1 and 2.2. The explicit specification of the relevant domains and (alleged) return mappings of levels 0 and 1 is contained in Appendices A.1, A.2 and A.3. For the proof of scaling, it remains only to verify, through straightforward application of the toolbox functions, that $\rho(0)$ and $\rho(1)$ are true first return maps with corresponding partitions of Ω, $D(0)$, and $D(1)$ which satisfy the renormalization equations. We have done this, obtaining affirmative outcomes for all relevant tests. The reader interested in the details is referred to the appendix of Kouptsov et al. (2002), as well as to the electronic supplement linked to that article.

For the strong recursive tiling proofs in Model I, the areas, weighted perimeters, and single-level return times (ν_j or $\tilde{\nu}_{\bar{j}}$) needed for application of Lemma 2.2 are collected in Table 2.3. In addition, with the help of the function `vertices(D)`, we

Table 2.3 Tiling data for Model I.

D	$\mathscr{A}(D)$	$\mathscr{W}(D)$	$\nu(D)$
$D_0(0)$	$-3-2\tau$	$4+3\tau$	6
$D_1(0)$	0	$-3-2\tau$	19
$D_2(0)$	$-4-\frac{5}{2}\tau$	$2+\frac{3}{2}\tau$	10
$D_3(0)$	0	$-3-2\tau$	27
$D_4(0)$	$\frac{13}{2}+4\tau$	$\frac{1}{2}+\frac{1}{2}\tau$	18
$D_0(1)$	$-21-13\tau$	$11+7\tau$	2
$D_1(1)$	0	$-8-5\tau$	4
$D_2(1)$	$-\frac{55}{2}-17\tau$	$\frac{11}{2}+\frac{7}{2}\tau$	4
$D_3(1)$	0	$-8-5\tau$	8
$D_4(1)$	$\frac{89}{2}+\frac{55}{2}\tau$	$\frac{3}{2}+\tau$	8

D	$\mathscr{A}(D)$	$\mathscr{W}(D)$	$\nu(D)$
$\Pi_0(\Omega)$	0	0	13
$\Pi_1(\Omega)$	0	$5+3\tau$	25
$\Pi_2(\Omega)$	0	0	17
$\Pi_3(\Omega)$	0	0	7
$\Pi_4(\Omega)$	0	0	1
$\Pi_5(\Omega)$	0	$2+\tau$	5
$\Pi_6(\Omega)$	$-\frac{29}{2}-9\tau$	$-\frac{25}{2}-\frac{15}{2}\tau$	4
$\Pi_0(0)$	0	$-3-2\tau$	2
$\Pi_1(0)$	0	0	2
$\Pi_2(0)$	0	$13+8\tau$	8
$\Pi_3(0)$	0	0	4
$\Pi_4(0)$	$-\frac{199}{2}-\frac{123}{2}\tau$	$-\frac{65}{2}-20\tau$	2
$\Pi_5(0)$	$-\frac{11}{2}-\frac{7}{2}\tau$	$\frac{15}{2}+5\tau$	1

find that there are 38 distinct vertices of the nondegenerate polygons in the partition of Ω which do not lie on excluded boundaries or on interior points of included segments. The analogous vertex count for $D(0)$ is 6. The verification of the four sum rules and two vertex counts is then completely straightforward.

In Model III, the strong recursive tiling is proved in similar fashion, making use of the data table A.3.6 in Appendix A. The weak recursive tiling property for Model II is simpler, requiring only verification of the area sum rules in Lemma 2.2.

2.5 Structure of the discontinuity set

2.5.1 Model I

An important byproduct of the strong recursive tiling of Ω is a comprehensive analysis of the structure of the discontinuity set Γ. For Model I, this structure is particularly simple: Γ is the union of the orbits of a countable collection of periodic domains, with a self-similar *baseline* on the boundary of Ω. As a preliminary to establishing this, we introduce the notion of *concatenation*.

Definition 2.4. Let A be an oriented line segment (open, closed, or half-open), and let A_i, $i = 1,\dots,n$ be a collection of disjoint oriented segments and points whose union is A. We say that A is the *concatenation* of $A_1,\dots,A_n$, written

$$A = A_1 \smile A_2 \smile \cdots \smile A_n,$$

if for each $i = 1,2,\dots,n-1$, the final boundary point of A_i coincides with the initial boundary point of A_{i+1}. By convention, the initial and final boundaries of a point coincide with the point itself.

We next prove the following lemma.

Lemma 2.3. *Every point in the discontinuity set Γ belongs to an orbit containing a point on the segment $\mathscr{B} = \Omega_6 \smile \Omega_8 = \{(x,y) : x = 0, -1 < y \leq 0\}$.*

Proof. The definition of Γ already establishes that all of its orbits intersect the included part of the boundary of Ω at least once. To restrict the latter to the segment $\mathscr{B}$, we easily verify:

$$\Omega_5 \subset K^{-1}(\Omega_6 \smile \Omega_9 \smile \Omega_7), \quad \Omega_7 \smile \Omega_9 \subset K^5(\Omega_6).$$

□

Our main result for the structure of Γ is then

Theorem 2.1. *Each point of Γ lies on a periodic orbit passing through a unique point of $\mathscr{B}$, the baseline. The latter is partitioned into the periodic points $P_n = (0, -|\lambda|^n)$, $n = 1,2,\dots$ and the periodic open segments $I_n = (P_n, P_n + 1)$, $n = 0,1,\dots$*

together with the fixed point at the origin. Each periodic point or segment lies on the return orbit of one of the scaling or pre-scaling domains, $\Pi_j(L)$ *or* $\Pi_j(\Omega)$. *Specifically,*

(A) $P_1, P_2, P_3, I_0, I_1, I_2$ *are, in order,* $\Pi_3(\Omega), \Pi_0(\Omega), \Pi_2(\Omega), \Pi_5(\Omega), K^8\Pi_0(0), \Pi_1(\Omega)$.
(B) *For* $L \geq 2$, P_{2L} *and* P_{2L+1} *lie on the orbits of* $\Pi_1(L-2)$ *resp.* $\Pi_3(L-2)$.
(C) *For* $L \geq 2$, I_{2L-1} *and* I_{Ln} *lie on the orbits of* $\Pi_0(L-1)$ *resp.* $\Pi_2(L-2)$.

Proof. Direct calculation establishes (A), as well as the decomposition

$$\mathscr{B} = (P_0, P_3] \smile K^{12}D_3(0) \smile (0,0).$$

From the latter, we see that $\mathscr{B}$ inherits the recursive tiling of $D_3(0)$. Specifically, for any level L,

$$D_3(L) = \rho(L)^2\Pi_0(L+1) \smile \Pi_1(L) \smile \Pi_2(L) \smile \Pi_3(L) \smile \rho(L)^4 D_3(L+1),$$

which follows from the data tables in Sect. A.1 of the Appendix. Statements (B) and (C) follow easily. The geometrical relations among the various points and segments is shown in Fig. 2.9. □

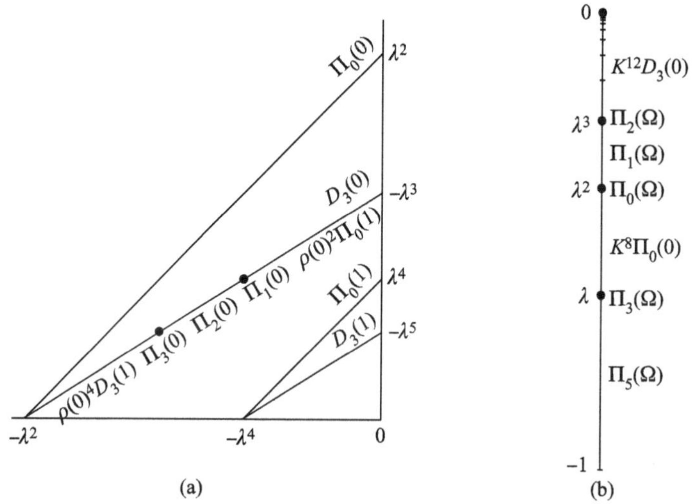

Fig. 2.9 Model I scaling domains **(a)** whose orbits intersect the discontinuity-set baseline $\mathscr{B}$, shown in **(b)**. The coordinates in (a) are relative to the scaling limit point at $(0, \tau)$.

2.5.2 *Model III*

For Model III, the discontinuity set has a much richer structure than that of Model I. In particular, Model I's discontinuity set contains no aperiodic orbits, that of Model III is host to an uncountable infinity of them (a Cantor set). The recursive inclusion structure of the baseline $\mathscr{B} = \Omega_7 \smile \Omega_8 \smile \Omega_9 \smile \Omega_{10}$ is laid out in great detail in Sect. A.3.7 of the Appendix. The data in this table were established rigorously by Kouptsov et al. (2002) using the same type of computer assisted proof used to prove the renormalizability of the model, making use of the same toolbox of elementary Mathematica functions. In the table, each segment of the baseline is listed as the image of some atom or periodic domain of the scaling sequence. From this information, one can easily infer the length and up/down orientation of any segment.

A notable feature of the table is the following: apart from a single image of the prescaling domain $\Pi_1(-1)$, it is only the scaling sequence $\Pi_3(L) \subset D_7(L)$, $L = 0,1,2,\ldots$ which participates in the recursive tiling. This does not prevent full coverage, up to a set of measure zero, of the baseline by periodic segments, a result which we now verify by explicit computation. At level $L \geq 0$, the baseline is tiled, up to a set of measure zero, by the periodic domains of level less than L and sets of $\nu_k(L)$ open segments congruent to $D_k(L)$, $L = 1,3,5,7$. From the table, one can easily read off the following recursion rules for determining $\nu_k(L)$:

$$\mathbf{v}(0) = \begin{pmatrix} \nu_1(0) \\ \nu_3(0) \\ \nu_5(0) \\ \nu_7(0) \end{pmatrix} = \begin{pmatrix} 6 \\ 9 \\ 9 \\ 6 \end{pmatrix}, \qquad \begin{pmatrix} \nu_1(L+1) \\ \nu_3(L+1) \\ \nu_5(L+1) \\ \nu_7(L+1) \end{pmatrix} = \begin{pmatrix} 1 & 0 & 2 & 1 \\ 2 & 1 & 2 & 1 \\ 3 & 1 & 2 & 0 \\ 1 & 1 & 1 & 1 \end{pmatrix} \begin{pmatrix} \nu_1(L) \\ \nu_3(L) \\ \nu_5(L) \\ \nu_7(L) \end{pmatrix}.$$

Since $\mathbf{v}(0)$ is an eigenvector of the recursion matrix with eigenvalue 5, we immediately get

$$\nu_7(L) = 6 \cdot 5^L, \qquad L = 0,1,2,\ldots.$$

Inserting (from the data tables)the lengths

$$|\Pi_3(L)| = \omega^{L+2}, \qquad |\Pi_1(-1)| = \omega, \qquad \omega = 3 - 2\sqrt{2},$$

we can sum up to get

$$\omega + 6\omega^2 \sum_{L=0}^{\infty} (5\omega)^L = \omega \left(\frac{1+\omega}{1-5\omega} \right) = \sqrt{2},$$

which is equal to the length of $\mathscr{B}$.

In Chapter 4, we will return to this model to study the fractal properties of the set $\mathscr{C}$ obtained by removing all of the periodic domains from $\mathscr{B}$. Then, in Chaps. 5 and 6, we will explore the remarkable long-time behavior of kicked oscillator orbits which originate in $\mathscr{C}$.

2.6 More general renormalization

For piecewise rational rotations of the square with quadratic irrational parameters, renormalization built around scaling and recursive tiling provides a comprehensive hierarchical framework for understanding the periodic and aperiodic orbits of the model. Going beyond the quadratic models requires an expanded concept of renormalization. Such a framework, again based on recursive tiling, was developed by Lowenstein et al. (2004) to handle the intricacies of the so-called $\pi/7$ model with cubic $\lambda = 2\cos\pi/7$ and will be the subject of the current section. How large a class of models can be accommodated by the formalism remains to be seen.

Before introducing the extended notion of recursive tiling, we first need to make more precise for the idea of *equivalence* of domains:

Definition 2.5. Two dressed domains $\Delta_1 = (D_1, \rho_1)$ and $\Delta_2 = (D_2, \rho_2)$ are equivalent, written $\Delta_1 \sim \Delta_2$, if they differ by a mapping $G \in \mathscr{G}$, where $\mathscr{G}$ is the group of geometric transformations defined in Sect. 2.2.3. Specifically,

$$\Delta_1 \sim \Delta_2 \Longleftrightarrow \exists G \in \mathscr{G},\ D_2 = GD_1,\ \rho_2 = G \circ \rho_1 \circ G^{-1}.$$

In our discussion of the quadratic models, the notion of equivalence entered implicitly, through the definition and renormalization properties of a scaling sequence. In all of the cases studied, the equivalence transformations were scale transformations with respect to a fixed point $\mathbf{x}_\infty$, involving only the subgroup generated by translations and scale transformations. In the more general scheme to be introduced below, we will remove this restriction and allow our equivalence relations to involve C-rotations and R-involution as well.

We now introduce our generalized recursion scheme. Suppose we have a *catalogue* of N inequivalent dressed domains, $\Delta^{[n]} = (D^{[n]}, \rho^{[n]})$ $n = 0, 1, \ldots, N$. Suppose further that each catalogue member has one or more dressed sub-domains $\Delta(i) = (D(i), \rho(i))$, $i = 0, 1, \ldots, I^{[n]}$, each of which is equivalent to some member of the catalogue. Note that the index i labels a link between parent and child dressed domains,

$$\Delta^{[n(i)]} \supset \Delta(i) \sim \Delta^{[h(i)]},$$

with $n(i)$ and $h(i)$ uniquely defined by these relations. Finally, suppose that each of the domains $D^{[n]}$ in the catalogue is tiled by a finite number of periodic domains together with the return orbits of the sub-domains $D_j(i)$, $n(i) = n, j = 0, 1, \ldots, J(h(i))$.

The catalogue members and their embedded sub-domains, each equivalent to a catalogue member, comprise the roots and first level of an infinite tree satisfying the recursive tiling property. Clearly, each $\Delta(i_0)$, being equivalent to $\Delta^{[h(i_0)]}$, has dressed sub-domains $\Delta(i_0, i_1)$, i_1 running over those index values such that $n(i_1) = h(i_0)$, and is tiled, up to periodic domains, by the return orbits of the various $D_{j_1}(i_0, i_1)$. Continuing in this fashion, one generates the entire hierarchy of level-L dressed domains $\Delta(\mathbf{i}) \stackrel{\text{def}}{=} \Delta(i_0, i_1, \ldots, i_L)$, where $i_k \in 0, 1, \ldots, I$ with the linkage $n(i_{k+1}) = h(i_k)$.

In Fig. 2.10, we represent schematically the tree generated by a hypothetical 3-member catalogue, each containing 2 dressed sub-domains equivalent to members of the catalogue. The tree is set beside the much simpler one of a quadratic model with a single scaling sequence. Note that in the more general framework, a scaling sequence can still make its a apearance, in the form of a path down the infinite tree in which all links are identical from some level onward. In this way, the quadratic models can be incorporated into the more general scheme, with the obvious changes of notation (e.g. $D_j(L)$ becomes $D_j(i_0,i_0,\ldots,i_0)$).

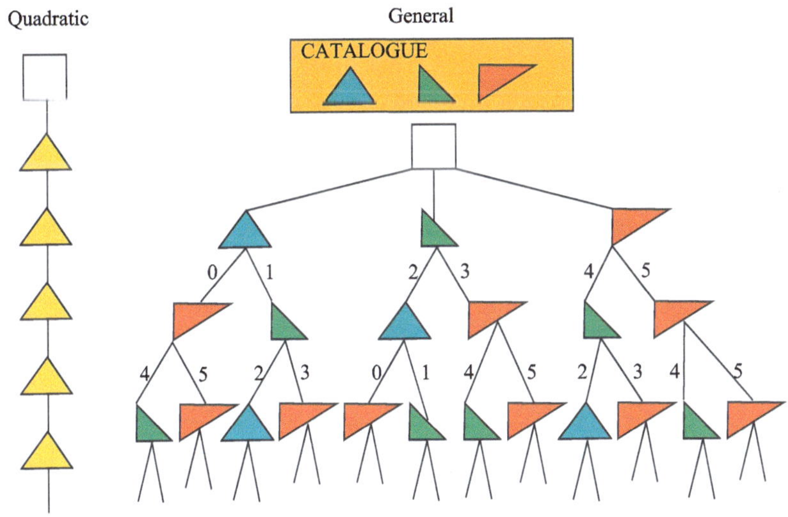

Fig. 2.10 Schematic representation of a fictitious 3-member catalogue and its recursive tree. At left, for comparison, is represented the scaling sequence of a typical quadratic model.

2.7 The $\pi/7$ model

We now focus on a nontrivial example of the more general recursive tiling, namely the $\pi/7$ model with its cubic irrational parameter $\lambda = 2\cos\pi/7$ (local rotation number $p/q = 1/14$) satisfying

$$\lambda^3 - \lambda^2 - 2\lambda + 1 = 0.$$

The map on $\Omega = [0,\tau)^2$, $\tau = \lambda^{-1}$ is shown in Fig. 2.11.

As a result of a strenuous, computer-aided search of the natural return map tree, it was discovered by Lowenstein et al. (2004) that a catalogue of four dressed triangles, $\Delta^{[n]}$, $n = 0,1,2,3$, is sufficient for a recursive tiling of Ω. The return-map partitions of these triangles are displayed in Figs. 2.12 and 2.13. There are 15 dressed sub-triangles whose return orbits implement the full tiling of the catalogue: $\Delta(9),\ldots,\Delta(14)$ tile $\Delta^{[0]}$, $\Delta(0),\ldots,\Delta(2)$ tile $\Delta^{[1]}$, $\Delta(3),\ldots,\Delta(5)$ tile $\Delta^{[2]}$, and $\Delta(6),\ldots,\Delta(8)$ tile $\Delta^{[3]}$. Each tiling covers the parent triangle up to a finite disjoint union of periodic cells, all of which are equivalent to a single prototype, the heptagon (notation of (2.5))

$$\Pi = \mathscr{D}\left((8,10,12,0,2,4,6),(1,\ldots,1),(0,\lambda,-1+\lambda+\lambda^2,\lambda+\lambda^2,-1+\lambda+\lambda^2,\lambda,0)\right). \tag{2.11}$$

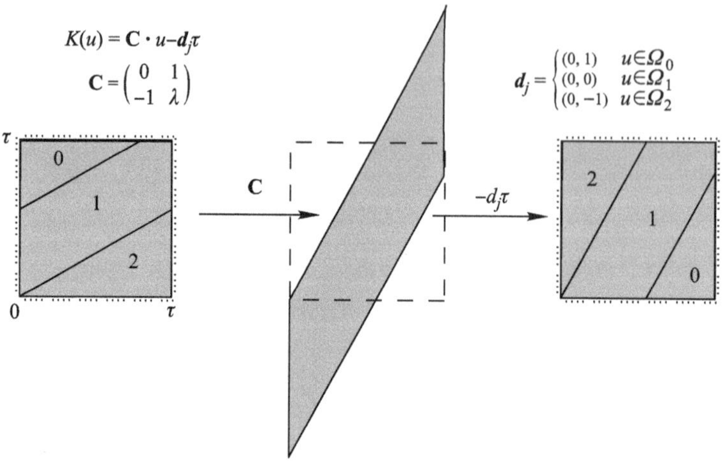

Fig. 2.11 Local map of the $\pi/7$ model.

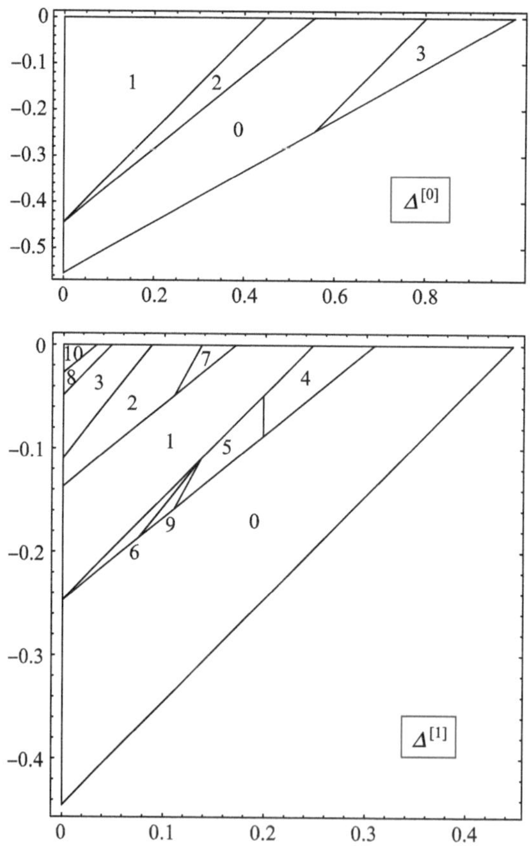

Fig. 2.12 Catalogue members $\Delta^{[0]}$ and $\Delta^{[1]}$ of the $\pi/7$ model, with return map partitions.

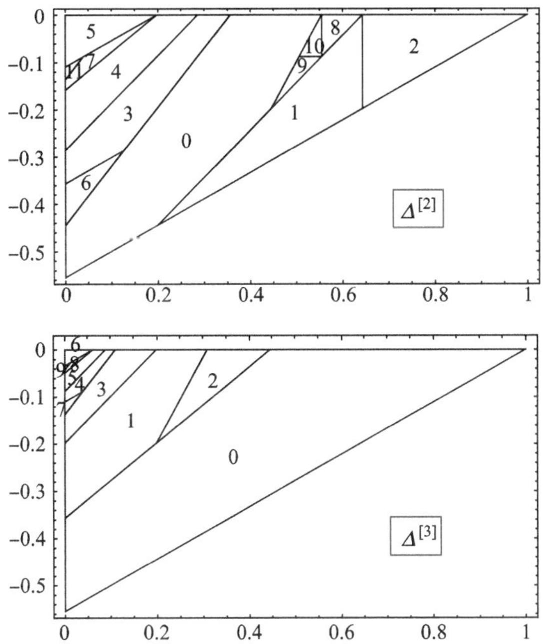

Fig. 2.13 Catalogue members $\Delta^{[2]}$ and $\Delta^{[3]}$ of the $\pi/7$ model, with return map partitions.

Recursion requires that each of the dressed sub-triangles $\Delta(i)$ be equivalent to a member of the catalogue. The correct parent-child relations, from which the infinite return map tree can be constructed recursively, are listed, together with other basic features of the catalogue members, in Table 2.4, and displayed graphically in Figs. 2.14, 2.15 and 2.16. Complete tables of geometrical and dynamical data needed for the computer assisted proof of renormalizability may be found in Sect. A.4 of the Appendix. Readers interested in the details of the proof itself can referred to Lowenstein et al. (2004) and its electronic supplement.

Table 2.4 Catalogue of prototype dressed domains $\delta^{[n]} = (D^{[n]}, \rho^{[n]})$, with $D^{[n]} = D(\mathbf{m}, (1,1,1), (0,0,1))$, with an overview of the recursive tree. Here $J^{[n]}$ is the number of atoms $D_j^{[n]}$ in the level-0 return map partition, and $\tilde{J}^{[n]}$ is the number of generating cells $\Pi_{\tilde{j}}^{[n]}$. The level-0 children of the prototypes are listed in the fourth column, together with the equivalence relations $\Delta(i_0) \sim \Delta^{[h(i_0)]}$.

$\Delta^{[n]}$	$\mathbf{m}$	$J^{[n]}$	Subdomains	$\tilde{J}^{[n]}$
$\Delta^{[0]}$	$(7,1,9)$	4	$\Delta(0), \Delta(2), \Delta(4) \sim \Delta^{[3]}$, $\Delta(1), \Delta(3), \Delta(5) \sim \Delta^{[1]}$	28
$\Delta^{[1]}$	$(7,1,11)$	11	$\Delta(6) \sim \Delta^{[2]}$, $\Delta(7), \Delta(8) \sim \Delta^{[1]}$	9
$\Delta^{[2]}$	$(7,1,9)$	12	$\Delta(9) \sim \Delta^{[2]}$, $\Delta(10), \Delta(11) \sim \Delta^{[1]}$	6
$\Delta^{[3]}$	$(7,1,9)$	10	$\Delta(12) \sim \Delta^{[3]}$, $\Delta(13), \Delta(14) \sim \Delta^{[1]}$	10

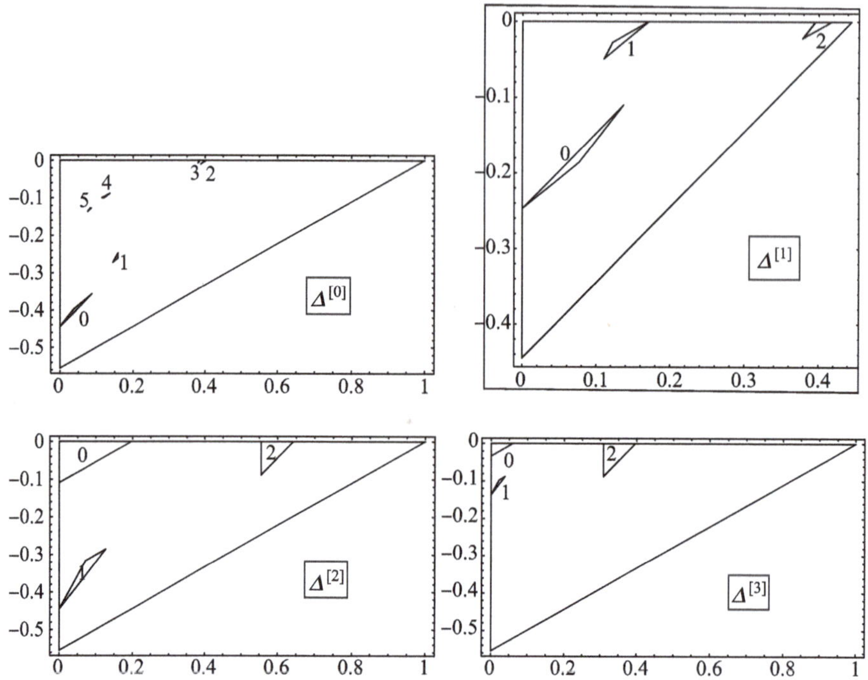

Fig. 2.14 Catalogue of the $\pi/7$ model, with dressed subdomains.

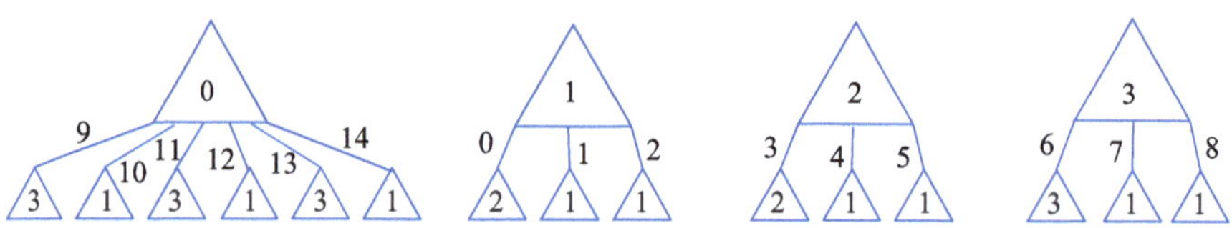

Fig. 2.15 Recursion rules for the catalogue of the $\pi/7$ model.

To begin the construction of the return map tree, we note that the first member of the catalogue, $\Delta^{[0]}$, is equivalent to the subdomain Ω_2 of the square. Figure 2.17 shows the tiling of the square by the 4 return map orbits of that dressed triangle. To proceed further down the tree, one recursively applies the 15 subdomain relations summarized in the graph of Fig. 2.16, in which the nodes are the catalogue members $\Delta^{[n]}$, $n = 0, \ldots, 3$, and the links are the 15 inclusion relations, with arrows pointing from parent to child. To each of these oriented links corresponds an embedding function $g(i)$, $i = 0, 1, \ldots, 14$, defined by (recall $h(i) =$ child, $n(i) =$ parent)

$$\Delta(i) = g(i)\Delta^{[h(i)]} \subset \Delta^{[n(i)]},$$
$$\Delta(\mathbf{i}_L) = \Delta(i_0, i_1, \ldots, i_L) = g(i_0, i_1, \ldots, i_L)\Delta^{[h(i_L)]} \subset \Delta(i_0, \ldots, \mathbf{i}_{L+1}),$$
$$g(\mathbf{i}_L) = g(i_0, \ldots, i_L) = g(\mathbf{i}_{L-1})g(i_L) = g(i_0)g(i_1)\cdots g(i_L).$$

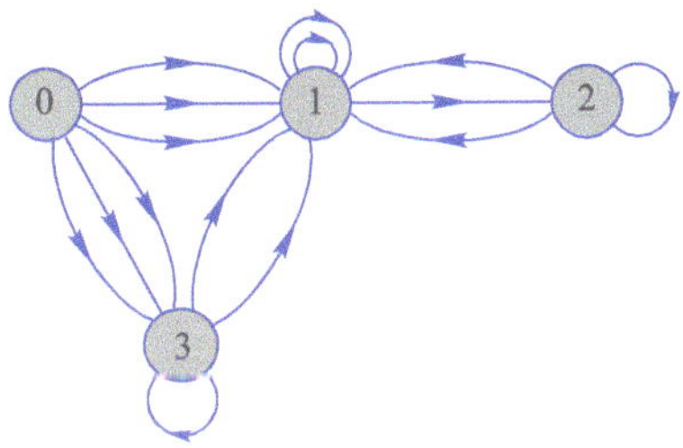

Fig. 2.16 Same as Fig. 2.15, in the form of a graph with oriented links.

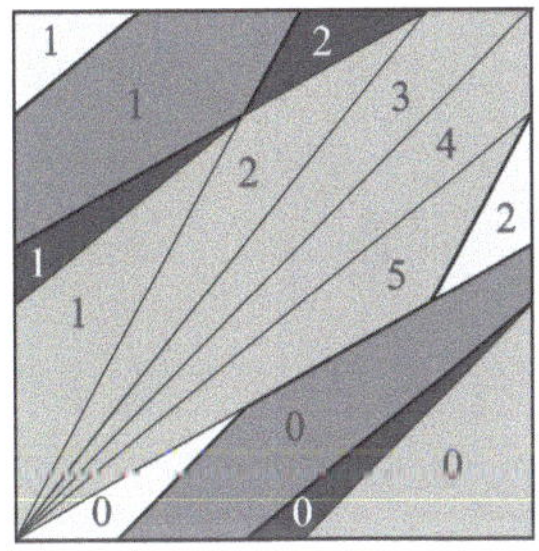

Fig. 2.17 Tiling of Ω by sub-domains of Ω_2.

As an illustration of the recursive tiling at level $L=0$, let us consider the tiling of $D^{[1]}$ by the return orbits of the $L=1$ subdomains $D_j(0),\ j=0,\ldots,11,\ D_j(1),\ j=0,\ldots,10$, and $D_j(2),\ j=0,\ldots,10$. Since

$$h(0)=2, \qquad h(1)=h(2)=1.$$

we only have to deal with the dressed triangles similar to $\Delta^{[1]}$ and $\Delta^{[2]}$ in this example. In order that the reader be able to follow the return orbits by eye, we have made a linear transformation to a coordinate system in which the metric is Euclidean, and the similarity transformations involve true rotations and reflections. The return map partitions and dressed subdomains for $\Delta^{[1]}$ and $\Delta^{[2]}$ are shown in Fig. 2.18, and the recursive tiling of $D^{[1]}$ is show, color coded, in Fig. 2.19.

In the recursive tiling, the basic building blocks of the level-L partition are the domains

$$D_{j_L}^{\mathbf{t}_L}(\mathbf{i}_L)=K^{t_0}\rho(i_0)^{t_1}\rho(i_0,i_1)^{t_2}\cdots\rho(\mathbf{i}_{L-1})^{t_L}D_{j_L}(\mathbf{i}_L),$$

which we shall refer to as *tiles*. An important consequence of the recursive tiling is that every point in the residual set Σ lies in the intersection of a nested sequence of tiles, i.e.

$$\{\mathbf{x}\}=\bigcap_L D_{j_L}^{\mathbf{t}_L}(\mathbf{i}_L),$$

and so $\mathbf{x}$ is uniquely determined by the infinite sequence

$$\sigma(\mathbf{x})=((i_0,j_0,t_0),(i_1,j_1,t_1),\ldots). \tag{2.12}$$

Note that once once j_L is known, all the preceding $j_{L'}$ are determined by the path constraints $j_k=p(i_{k+1},j_{k+1},t_{k+1})$.

The set $\Sigma(\mathbf{i}_L)=\bigcup_{j_L,\mathbf{t}_L}D_{j_L}^{\mathbf{t}_L}(\mathbf{i}_L)\cap\Sigma$ consists of all discontinuity-avoiding orbits which pass through $D(\mathbf{i}_L)$ and is clearly an *invariant* subset of Σ. This property extends to the intersection $\Sigma(\mathbf{i})=\bigcap_L\Sigma(\mathbf{i}_L)$, for any infinite linked sequence $i_0,i_1,\ldots,\ \forall L,h(i_L)=n(i_{L+1})$, i.e. to any path down the infinite tree generated by the recursion rules of the catalogue.

The existence of a multiplicity of invariant sets implies a profound contrast between the quadratic models of Sect. 3, in which the recursion trees are simple lad-

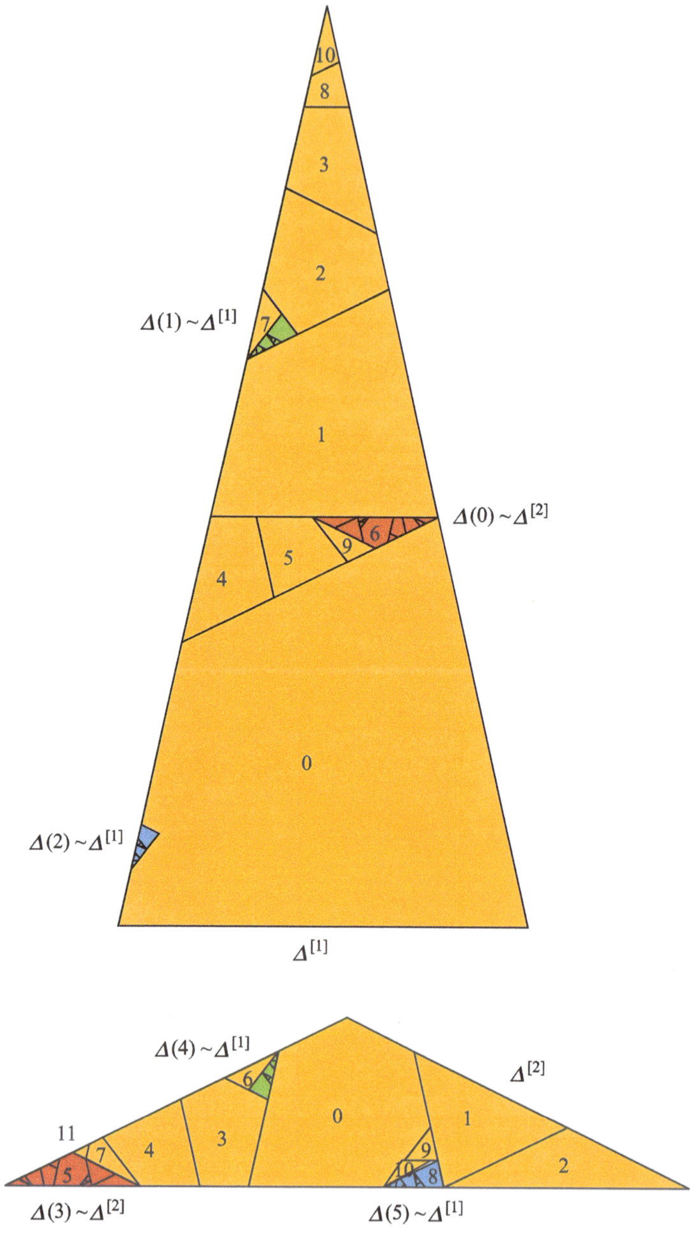

Fig. 2.18 Return map partitions of catalogue members $D^{[1]}$ and $D^{[2]}$ into their respective atoms. The Euclidean metric is used here to make evident the various similarity relations. Superposed are the dressed subdomains $\Delta(i)$, $i = 0, \ldots, 5$. (See the color figure at the end of the book.)

Fig. 2.19 Tiling of $D^{[1]}$ by the return orbits of $D_j(i)$, $i = 0, 1, 2$, $j = 0, \ldots, J(h(i))$. The Euclidean metric is used to make evident the similarity relations. Consistent with the previous figure, different values $i = 0, 1, 2$ are coded red, green, and blue, respectively. (See the color figure at the end of the book.)

ders, and models with more elaborate trees, such as the $\pi/7$ model. In the latter, there are, except for level 0, where $i_0 = 0$, and level 1, where $i_1 \in \{9, 10, \ldots, 14\}$, exactly 3 alternatives at each branching of the tree. Thus, the number of branches at level $L \geq 1$ is $6 \times 3^{L-1}$, which for $L \to \infty$ tends to an uncountable infinity of alternatives. To see this, we can map the sequences $i_2, i_3, \ldots$ onto the set of infinite ternary sequences, hence onto the continuum interval $[0,1]$.

The phase portrait of K on the residual set thus splits up into sectors labeled by the code sequences $\mathbf{i}$. Each sector has a dynamics which parallels, to a large extent, a typical quadratic model with a single scaling sequence. For example, in the symbolic representation (2.12), the indices i_k are an inert background, relevant only through the path constraint $j_k = p(i_{k+1}, j_{k+1}, t_{k+1})$. All of the points on the forward orbit of the point x have the same sector indices $i_0, i_1, \ldots$. Within a sector, the level-L return times are determined by iteration of the incidence matrices

$$A_{j,k}(i) = \sum_{t=0}^{v_k - 1} \delta_{j,p(i,k,t)}, \quad j,k = 0, \ldots, J(h(i)).$$

$$\boldsymbol{A}(\mathbf{i}_L) = \boldsymbol{A}(\mathbf{i}_{L-1}) \cdot \boldsymbol{A}(i_L) = \boldsymbol{A}(i_0) \cdot \boldsymbol{A}(i_1) \cdots \boldsymbol{A}(i_L),$$

$$T_k(i) = \sum_{j=0}^{J(h(i)} A_{j,k}(i), \quad k = 0, \ldots, J(h(i)),$$

$$\boldsymbol{T}(\mathbf{i}_L) = (T_0(\mathbf{i}_L), \ldots, T_{J(h(i_L))}(\mathbf{i}_L)) = \boldsymbol{T}(\mathbf{i}_{L-1}) \cdot \boldsymbol{A}(i_L) = \boldsymbol{T}(i_0) \cdot \boldsymbol{A}(i_1, \ldots, i_L). \quad (2.13)$$

The level-to-level scale factors are also determined recursively,

$$\omega(\mathbf{i}_L) = \omega(\mathbf{i}_{L-1})\omega(i_L) = \omega(i_0)\omega(i_1)\cdots\omega(i_L).$$

As in the quadratic models, the scale factors are all *units*. As pointed out in Section 2.2.1, these are all of the form

$$\eta_1^{m_1}\eta_2^{m_2},$$

where η_1 and η_2 are so-called fundamental units. For convenience, we choose these to be

$$\eta_1 = 2 + \lambda - \lambda^2 = \tau = \lambda^{-1}, \qquad \eta_2 = -1 - \lambda + \lambda^2 = (-1+\lambda)^{-1} = 1 - \eta_1. \quad (2.14)$$

The fact that λ and τ are units is a special feature of this model, not necessarily expected to hold for more general parameter values.

For almost all sectors i, the domains $D(\mathbf{i}_L)$, $L = 0, 1, \ldots$, do not form a scaling sequence, since the shapes and return map dynamics keep changing from one level to the next. However, among the various sectors, there are indeed infinitely many scaling sequences, namely those for which $\mathbf{i}$ is eventually periodic, i.e. of the form

$$\ldots i_a, \ldots, i_b, i_a, \ldots, i_b, i_a, \ldots, i_b, \ldots = \ldots (i_a, \ldots, i_b)^{\infty}.$$

In that case we have a scaling sequence whose limit point is $\mathbf{x}_\infty$ with code sequence $\sigma(\mathbf{x}_\infty) = \mathbf{i}$. Its temporal scale factor $\omega_T(\mathbf{i})$ is equal to the largest eigenvalue of $(A(i_a)\cdots A(i_b))^T$, and its geometric scale factor is $\omega_K(\mathbf{i}) = \omega(i_a)\cdots\omega(i_b)$. By the informal argument at the end of Sect. 2.1, the set $\Sigma(\mathbf{i})$ is expected to be a fractal of dimension $-\log\omega_T(\mathbf{i})/\log\omega_K(\mathbf{i})$. A more precise discussion will be given in Chapter 4.

Although the countable set of eventually periodic sector codes is of negligible size within the uncountable set of all codes, it is nevertheless of considerable importance, analogous to that of the rational numbers among the real's. Just as the rational numbers are essential for practical numerical computation, the sectors with eventually periodic codes are essential for the explicit construction of orbits in the residual set.

References

Adler R., Kitchens B., and Tresser C. (2001) Dynamics of nonergodic piecewise affine maps of the torus, *Ergodic Theory and Dynamical Systems* **21**, 959–999.

Akiyama S., Brunotte H., Pethö A., and Steiner W. (2008) Periodicity of certain piecewise affine planar maps, *Tsukuba Journal of Mathematics* **32**, 1–55.

Ashwin P. (1997) Elliptic behaviour in the sawtooth standard map, *Physics Letters A* **232**, 409–416.

Ashwin P., Chambers W., and Petrov G. (1997) Lossless digital filter overflow oscillations: approximations of invariant fractals, *International Journal of Bifurcation and Chaos* **7**, 2603–2610.

Boshernitzan M.D. and Carroll C. R. (1997) An extension of Lagrange's theorem, *Journal d'Analyse Mathématique* **72**, 21–44.

Buzzi J. (2001) Piecewise isometries have zero topological entropy, *Ergodic Theory and Dynamical Systems* **21**, 1371–1377.

Chua L.O. and Lin T. (1988) Chaos in digital filters, *IEEE Transactions:Circuits and Systems* **CAS-35**, 648–658.

Chua L.O. and Lin T. (1990) Fractal pattern of second order non-linear digital filters: a new symbolic analysis, *International Journal of Circuit Theory and Applications* **18**, 541–550.

Dana, I. (2004) Global superdiffusion of weak chaos, *Physical Review E* **69**, 016212.

Davies A.C. (1995) Nonlinear oscillations and chaos from digital filters overflow, *Philosophical Transactions of the Royal Society London* **A 353**, 85–99.

Falconer K. (1990) *Fractal Geometry*, Wiley, Chichester.

Goetz A. (1996) *Dynamics of piecewise isometries*, PhD Thesis, University of Chicago.

Goetz A. (1998) Dynamics of a piecewise rotation, *Continuous and Discrete Dynamical Systems* **4**, 593–608.

Goetz A. (2000) Dynamics of piecewise isometries, *Illinois Journal of Mathematics* **44**, 465–478.

Goetz A. (2001) Stability of cells in non-hyperbolic piecewise affine maps and piecewise rotations, *Nonlinearity* **14**, 205–219.

Goetz A. and Poggiaspalla G. (2004) Rotation by $\pi/7$, *Nonlinearity* **17**, 1787–1802.

G.H. Hardy and E.M. Wright (1979) *An Introduction to the Theory of Numbers*, 5th edition, Oxford University Press, Oxford.

Kahng B. (2000) Dynamics of symplectic affine maps on tori, PhD Thesis, University of Illinois at Urbana-Champaign.

Kahng B. (2002) Dynamics of symplectic piecewise affine elliptic rotation maps on tori, *Ergodic Theory and Dynamical Systems* **22**, 483–505.

Khinchin A. Ya. (1964) *Continued Fractions*, Dover, Mineola, New York.

Kouptsov K.L., Lowenstein J.H., and Vivaldi F. (2002) Quadratic rational rotations of the torus and dual lattice maps, *Nonlinearity* **15**, 1795–1842.

Lowenstein J.H. (2007) Aperiodic orbits of piecewise rational rotations of convex polygons with recursive tiling, *Dynamical Systems* **22**, 25–63.

Lowenstein J.H. and Vivaldi F. (2000) Embedding dynamics for round-off errors near a periodic orbit, *Chaos* **10**, 747–755.

Lowenstein J.H., Hatjispyros S., and Vivaldi, F. (1997) Quasi-periodicity, global stability and scaling in a model of Hamiltonian round-off, *Chaos* **7**, 49–66.

Lowenstein J.H., Kouptsov K. L., and Vivaldi F. (2004) Recursive tiling and geometry of piecewise rotations by $\pi/7$ *Nonlinearity* **17**, 1–25.

Lyubomudrov O., Edelman M., and Zaslavsky G. M. (2003) Pseudochaotic systems and their fractional kinetics, *Int. Journal of Modern Physics B* **17**, 4149–4167.

Poggiaspalla G. (2006) Self-similarity in piecewise isometric systems, *Dynamical Systems* **21**, 147–189.

Zaslavsky G. M. and Edelman M. (2003) Pseudochaos, *Perspectives and Problems in Nonlinear Science: a Celebratory Volume in Honor of Lawrence Sirovich*, eds. E. Kaplan, J. Marsden, and K. R. Sreenivasan, Springer, New York, 421–423.

Wu C.W. and Chua L.O. (1993) Properties of admissible sequences in a second order digital filter with overflow non-linearity, *International Journal of Circuit Theory and Applications* **21**, 299–307.

Chapter 3
Symbolic Dynamics

In the preceding chapter, we introduced a hierarchy of covering sets (recursive tilings) which converge to the residual set of aperiodic, discontinuity-avoiding orbits of the local map. We now want to encode the points of the residual set abstractly as strings of symbols (codes), in such a way that application of the dynamical map is conjugated to a relatively simple transformation of the codes, implemented by the odometer-like machine which we will call a *codometer*.

One of the themes which we want to emphasize in this book is the fact that the kicked oscillator maps and the corresponding piecewise isometries on the square may be considered as two different representations of the same symbolic dynamics. In some cases, we will even have a third representation, in the form of a model of Hamiltonian round-off (see Chapter 7). For the symbolic dynamics to be a meaningful underpinning of these quite different geometric mappings, it must be defined intrinsically. That is, there must be a set of pruning rules for distinguishing the admissible codes from the inadmissible ones. Much of the current chapter will be devoted to this nontrivial task.

3.1 Symbolic representation of the residual set

There is a natural symbolic representation of points in the residual set $\Sigma^{[n]}$, i.e. points which lie neither in periodic cells nor on discontinuity lines of a catalogue member $\Delta^{[n]} = (D^{[n]}, \rho^{[n]})$. Let us introduce this coding, step by step, for a given point $\mathbf{x}$.

3.1.1 Hierarchical symbol strings

Thanks to the recursive tiling of $D^{[n]}$ by level-0 tiles $D^{t_0}_{j_0}(i_0)$, $j_0 = 0, 1, \ldots, J^{[h(i_0)]}$, $t_0 = 0, 1, \ldots, \nu_j(i_0)$, we can identify the unique tile within which $\mathbf{x}$ resides, assigning it the symbol (i_0, j_0, t_0). Since $\mathbf{x}$ is assumed not to lie on any tile boundary, we can work here, and in the rest of this chapter, entirely with open polygonal tiles, making use only of the weak version of recursive tiling.

To pin down the location of $\mathbf{x}$ with higher resolution, we can go to level 1 and locate the tile

$$D^{t_0t_1}_{j_1}(i_0i_1) \subset D^{t_0}_{j_0}(i_0)$$

which (again uniquely) contains $\mathbf{x}$, adding a second symbol to the list,

$$\mathbf{x} \leftrightarrow (i_0, j_0, t_0), (i_1, j_1, t_1), \ldots \tag{3.1}$$

indicating with dots the additional symbols to be specified as we proceed. Note that one of the component symbols, j_0, is redundant in the notation, due to the path constraint built into the renormalization structure:

$$j_0 = p(i_1, j_1, t_1).$$

Moreover, the indices i_0 and i_1 are not completely independent: recall that in general the index i stands for an inclusion of one dressed domain, equivalent to catalogue member $D^{[h(i)]}$, within another, equivalent to the catalogue member $D^{[n(i)]}$, and so in (3.1) we must have

$$h(i_0) = n(i_1).$$

Rather than eliminating such redundancies in the notation, we will retain them as valuable signposts of the recursive structure.

Clearly the above method allows us to construct as small a neighborhood of the target point $\mathbf{x}$ as we want, simply by finding a sufficiently small tile of the form

$$D^{t_0\ldots t_L}_{j_L}(i_0 \ldots i_L) \subset D^{t_0\ldots t_{L-1}}_{j_{L-1}}(i_0 \cdots i_{L-1}) \subset \cdots \subset D^{t_0}_{j_0}(i_0).$$

The recursive tiling property ensures that such a tile always exists. The specification of all such tiles for $L = 0, 1, \ldots, \infty$ pins down the point $\mathbf{x}$ exactly, and so we write the correspondence in the form

$$\mathbf{x} \leftrightarrow ((i_0, j_0, t_0), (i_1, j_1, t_1), \ldots, (i_k, j_k, t_k), \ldots), \tag{3.2}$$

$$h(i_k) = n(i_{k+1}), \qquad j_k = p(i_{k+1}, j_{k+1}, t_{k+1}), \qquad k = 0, 1, 2, \ldots. \tag{3.3}$$

A constrained symbol sequence of the form (3.2), with (3.3), will be called a *linked symbol sequence*, or, more simply, a *code*. Although each $\mathbf{x} \in \Sigma^{[n]}$ determines a unique linked sequence, the reverse is not always the case: a linked symbol sequence might instead determine a point on one of the boundary lines (since an infinite sequence of nested domains can have as their limit a point on a bounding discontinuity line). The linked sequences which correspond to points in $\Sigma^{[n]}$ will be called *admissible*, while those which relate to points on boundary lines will be called *inadmissible*. To meaningfully exploit the advantages of symbolic dynamics, we will need a way to distinguish *intrinsically* the admissible from the inadmissible codes. This task will occupy our efforts in Sect. 3.3.

3.1.2 Eventually periodic codes

Among the admissible linked symbol sequences, an important subset is that of the eventually periodic ones. Like the rational numbers embedded in the reals, the points corresponding to eventually periodic codes form a countably infinite dense skeleton within the residual set. It is these points which we normally deal with in computational experiments, primarily because they can be located precisely, as algebraic numbers. Thus, even though $\Sigma^{[n]}$ is a set of Lebesgue measure zero so that the problem of picking out a sample point by trial and error is zero, we can nonetheless find plenty of $\mathbf{x} \in \Sigma^{[n]}$ by exploiting their eventually periodic codes.

To see how the construction works, let us suppose that we are given a period-L linked symbol sequence,

$$\sigma = (\sigma_1, \sigma_2, \ldots, \sigma_L)^\infty, \; \sigma_k = (i_k, j_k, t_k),$$

so that

$$\mathbf{x} = \lim_{N\to\infty} \rho^{[n(i_1)]t_1} \rho(i_1)^{t_1} \rho(i_1, i_2)^{t_2} \cdots \rho(\mathbf{i}_{NL-1})^{t_{NL}} D^{\mathbf{0}}_{j_{NL}}(\mathbf{i}_{NL}).$$

where

$$D^{\mathbf{0}_k}_{j_k}(\mathbf{i}_k) \stackrel{\text{def}}{=} D^{00\cdots 0}_{j_k}(i_0, \ldots, i_k).$$

The recursive tiling allows us to write

$$D^0_j(i) = g(i) D^{[h(i)]}_j, \quad \rho(i) = g(i) \rho^{[n]} g(i)^{-1},$$

where $g(i) \in \mathscr{G}$, and more generally

$$D^{\mathbf{0}_k}_j(i_1, \ldots, i_k) = g(\mathbf{i}_k) D^{[h(i_k)]}, \quad \rho(\mathbf{i}_k) = g(\mathbf{i}_k) \rho^{[h(i_k)]} g(\mathbf{i}_k)^{-1},$$

where $g(\mathbf{i}_k) = g(i_1) g(i_2) \cdots g(i_k)$. Thus

$$\mathbf{x}(\sigma) = \lim_{N\to\infty} g^N D^0_{j_{NL}}(i_{NL}), \tag{3.4}$$

where

$$g = \rho^{[n(i_1)]t_1} g(i_1) \rho^{t_2}(i_1) g(i_2) \cdots \rho^{t_L}(i_{L-1}) g(i_L).$$

Using the group commutation relations, $g \in \mathscr{G}$ can be expanded as

$$g = T_\delta R^\alpha C^\nu S_\omega = T_\delta g',$$

where, since $0 < \omega < 1$, g' is a strictly contracting linear transformation. Thus, from (3.4), we have

$$\mathbf{x}(\sigma) = g\mathbf{x}(\sigma) = g'\mathbf{x}(\sigma) + g\mathbf{0}, \tag{3.5}$$

where $\mathbf{0}$ is the origin, a fixed point of g'. Since $1 - g'$ is invertible,

$$\mathbf{x}(\sigma) = (1 - g')^{-1} g\mathbf{0}.$$

This provides a straightforward algorithm for calculating $\mathbf{x}(\sigma)$ for repeating σ. For an eventually periodic sequence, a finite number of additional group premultiplications are sufficient to yield the result.

3.1.3 Simplified codes for quadratic models

For the quadratic models we deal with in this book, the linked index sequences $\mathbf{i} = i_0, i_1, i_2, \ldots$ which appear in the coding are exceedingly simple. Typically, one has

$$\mathbf{i} = i_0, i_1, i_1, i_1, \ldots \stackrel{\text{def}}{=} i_0, i_1^\infty,$$

where i_0 corresponds to the inclusion of the base domain of a scaling sequence in the generating dressed domain (Ω, K), and i_1 corresponds to the inclusion of one member of a scaling sequence within another. The most complicated of the quadratic examples introduced in Chapter 2 was Model III, where the multiindices would be of the form

$$\mathbf{i} = i_0, i_1, i_2, i_3, i_2, i_3, \ldots = i_0, i_1, (i_2, i_3)^\infty,$$

with the respective i_k corresponding to the inclusions

$$D(-1) \subset \Omega, \qquad D(0) \subset D(-1), \qquad D(k \text{ odd}) \subset D(k-1 \text{ even}),$$
$$D(k \text{ even}) \subset D(k-1 \text{ odd}).$$

Note that if we are only interested in points of the residual set, the distinction between even and odd levels disappears, and so, with telescoping, we can use multiindices which conform to the simpler pattern of the other examples.

In most contexts, the multiindex $\mathbf{i}$ is unique and unchanging, and so it makes sense in the quadratic models to suppress it in the symbolic representation, writing

$$\sigma(\mathbf{x}) = (j_0, t_0)(j_1, t_1)(j_2, t_2) \cdots .$$

We will adopt this convention in what follows.

3.2 Dynamical updating of codes

Having introduced a representation of each point $\mathbf{x}$ in $\Sigma^{[n]}$ as an infinite symbol string σ, we now want to determine how σ is transformed when $\mathbf{x}$ is mapped into $\rho^{[n]}(\mathbf{x})$. We begin by focusing on the first symbol in the string, (i_0, j_0, t_0), which assigns $\mathbf{x}$ to the level-0 tile

$$D_{j_0}^{t_0}(i_0) = \rho^{[n] t_0} D_{j_0}(i_0), \quad 0 \le t_0 \le \nu_{j_0}(i_0) - 1 \stackrel{\text{def}}{=} t_0^{\max}$$

Clearly, if t_0 is not maximal, i.e. if $t_0 < \nu_{j_0}(i_0) - 1$, then applying $\rho^{[n]}$ maps $D_{j_0}^{t_0}(i_0)$ to $D_{j_0}^{t_0+1}(i_0)$, and hence

$$(i_0, j_0, t_0 < t_0^{\max}) \cdots \mapsto (i_0, j_0, t_0 + 1) \cdots,$$

where the dots correspond to an arbitrary admissible continuation to the right, the same before and after the mapping. On the other hand, if t_0 is maximal, we have

$$\rho^{[n]} D_{j_0}^{t_0^{\max}}(i_0) = \rho^{[n]\,\nu_{j_0}(i_0)} D_{j_0}(i_0) = \rho(i_0) D_{j_0}(i_0).$$

Thus, the application of $\rho^{[n]}$ completes the level-0 return orbit of $D_{j_0}(i_0)$, bringing it back into $D(i_0)$. Here we are faced with the possibility of branching, since $\rho(i_0)$ acts piecewise on the level-1 domains which tile $D(i_0)$. The outcome is, of course, determined by bringing into play the second symbol in the representation of $\mathbf{x}$. Specifically, if

$$\sigma = (i_0, j_0, t_0^{\max})(i_1, j_1, t_1) \cdots,$$

where t_1 is submaximal, we have

$$\mathbf{x} \in D_{j_1}^{t_0^{\max} t_1}(i_0, i_1),$$
$$\rho^{[n]} \mathbf{x} \in \rho(i_0) D_{j_1}^{0 t_1}(i_0, i_1) = D_{j_1}^{0 t_1 + 1}.$$

Thus, the level-1 return orbit of the base tile $D_{j_1}(i_0, i_1)$ advances by one step. The level-0 atom within which the image tile lies is determined by the path function:

$$D^{0 t_1 + 1}(i_0, i_1)_{j_1} \subset D_{p(i_1, j_1, t_1 + 1)}(i_0),$$

and so we get the symbolic mapping

$$(i_0, j_0, t_0^{\max})(i_1, j_1, t_1 < t_1^{\max}) \cdots \mapsto (i_0, p(i_1, j_1, t_1 + 1), 0)(i_1, j_1, t_1 + 1) \cdots, \quad (3.6)$$

where once again the dots indicate the same admissible extensions on both sides of the arrow.

Clearly the process can be iterated ad infinitum to obtain the general rule

$$\begin{aligned}(i_0, j_0, t_0^{\max})(i_1, j_1, t_1^{\max}) \cdots (i_k, j_k, t_k < t_k^{\max})(i_{k+1}, j_{k+1}, t_{k+1} < t_{k+1}^{\max}) \cdots \\ \mapsto (i_0, j_0', 0)(i_1, j_1', 0) \cdots (i_k, j_k', 0)(i_{k+1}, j_{k+1}, t_{k+1} + 1) \cdots\end{aligned} \quad (3.7)$$

$$j_r' = p(i_{r+1}, j_{r+1}', 0), \; r = 0, \ldots, k-1, \qquad j_k' = p(i_{k+1}, j_{k+1}, t_{k+1} + 1).$$

This leaves only the case where *all* t_k are maximal, which, according to the preceding pattern, should transform as

$$(i_0, j_0, t_0^{\max})(i_1, j_1, t_1^{\max}) \cdots (i_k, j_k, t_k^{\max}) \cdots \mapsto (i_0, j_0', 0)(i_1, j_1', 0) \cdots (i_k, j_k', 0) \cdots \quad (3.8)$$

$$j'_k = p(i_{k+1}, j'_{k+1}, 0),\ k = 0, 1, 2, \ldots.$$

At first sight, this prescription appears to be non-unique. Fortunately, we have constructed all of our models in such a way that each base domain $D(\mathbf{i}_{L+1})$ is contained in a single atom of $D(\mathbf{i}_L)$, and so j'_k depends only on i_{k+1}, not on j'_{k+1}.

The symbolic representation of the dynamics makes very clear the invariance of each set $\Sigma(i_0, i_1, \ldots)$ consisting of all those points of the residual set which share the same linked index sequence $i_0, i_1, \ldots$, $h(i_k) = n(i_{k+1})$. On each of the $\Sigma(\mathbf{i})$, the hierarchical coding and updating scheme just described is an example of a Vershik map on an ordered Bratteli diagram (Bratteli, 1972; Vershik,1985; Durand et al., 1999). This association was discussed by Poggiaspalla (2006) in the case of periodic sequences $(i_1, i_2, \ldots, i_L)^\infty$, where the Bratteli diagram is stationary, and generalized to the present context by Lowenstein (2007). Simple examples of such a dynamical map are the *simple odometer*, where all i_k and j_k can be set to zero and the code reduces to $(t_1, t_2, \ldots)$, and the *multi-register odometer*, with code $((i_0, t_0), (i_1, t_1), \ldots)$, in which each level has a fixed identity i_k with $0 \le t_k < \nu(i_k)$. Dynamical mappings of the odometer type were used by Afraimovich et al. (2000) to model sticky orbits in Hamiltonian systems.

The hierarchical updating scheme defined above is conveniently implemented as a symbolic machine, the *codometer*. The latter is described in elementary terms in Appendix B, with application to a familiar example, the Gregorian calendar.

3.3 Admissibility

Thus far we do not have a completely satisfactory symbolic dynamics, thanks to the problem of admissibility, i.e. distinguishing those sequences which correspond to points of Σ from those which correspond to points of $\overline{\Sigma} \setminus \Sigma$. Geometrically, the distinction is not difficult to specify. In each case, the code $\sigma(x) = (\sigma_0, \sigma_1, \ldots)$ corresponds to a nested sequence of open domains $D_0 \supset D_1 \supset \cdots$, with $x = \lim_{L \to \infty} D_L$. Each D_L is in fact a tile $D_{j_L}^{t_0,\ldots,t_L}(i_0, \ldots, i_L)$, but for the time being we dispense with the detailed indices. For $x \in \Sigma$, x lies in the interior of each D_L by definition, while for $y \in \overline{\Sigma} \setminus \Sigma$, y lies on the boundary ∂D_L of at least one D_L. In fact, if $y \in \partial D_L$, y also lies on ∂D_{L+1}, and hence is a boundary point of all D_K, $K \ge L$.

The inadmissible sequences thus correspond to the common boundary points of infinite nested sequences of tiles. We now wish to associate these points with infinite sequences of tile edges, the latter being defined to include their endpoints, the tile vertices. Suppose that x is not a vertex of any of the domains associated with the code $\sigma(x)$. Then it is clear that x lies in the intersection of a nested set of tile edges. If, on the other hand, x is a vertex of D_L, it is not obvious that x is the endpoint of a nested set of edges $E_{L'}$, $L' \ge L$. In particular, $x \in E_L$ does not imply that there exists a matching edge E_{L+1} of D_{L+1}, as one sees in Fig. 3.1. We note that this can happen only if the vertex angle at x for D_{L+1} is *strictly less* than that for D_L. But in our models, only finitely many vertex angles are possible, and so there always exists an $L_{\min}$ and a nested set of edges $E_L \subset \partial D_L$, with $x \in E_L, L \geq L_{\min}$.

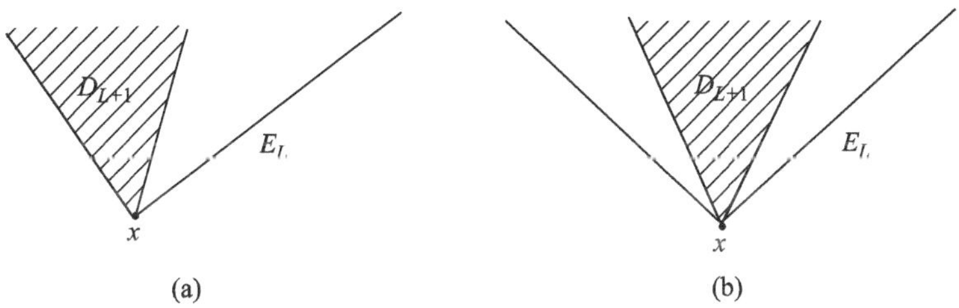

Fig. 3.1 Matching vertices for D_L and D_{L+1}, with **(a)** and without **(b)** matching edges.

Recursive edge-matching is thus the key to checking the tails of symbol sequences for admissibility. Fortunately, our description of polygonal domains in terms of half-plane inequalities in Sect. 2.2.2 allows for a systematic enumeration of edges which is level-independent and thus well suited to testing for matching edges. Specifically, the a'th edge of $\mathscr{D}(\mathbf{m}', \mathbf{s}', \mathbf{b}')$ matches the ath edge of $\mathscr{D}(\mathbf{m}, \mathbf{s}, \mathbf{b})$ if

$$m'_{a'} = m_a, \quad s'_{a'} = s_a, \quad b'_{a'} = b_a, \tag{3.9}$$

or, if the parameter λ is equal to $2\cos(2\pi p/q)$ with q even,

$$m'_{a'} = m_a + q/2 \bmod q, \quad s'_{a'} = -s_a, \quad b'_{a'} = -b_a. \tag{3.10}$$

None of the operations of generalized rotation, reflection, translation, or scaling which generate the transformation group $\mathscr{G}$ affects the order of the indices in $\mathbf{m}, \mathbf{s}, \mathbf{b}$, and hence the edge enumeration is $\mathscr{G}$-invariant. Since the return maps relating different levels are elements of $\mathscr{G}$, the level-independence of the edge assignments follows.

We now have an abstract criterion for inadmissibility, which we need to turn into a simple rule for testing specific symbol sequences. We already have such a rule for the linked sequences: given the kth symbol (i_k, j_k, t_k), the $(k+1)$st symbol $(i_{k+1}, j_{k+1}, t_{k+1})$ must satisfy

$$n(i_{k+1}) = h(i_k), \quad p(i_{k+1}, j_{k+1}, t_{k+1}) = j_k, \tag{3.11}$$

but is otherwise arbitrary. This rule can be conveniently represented by a transition graph or equivalent transition matrix of 0's and 1's. An inadmissible tail sequence has in addition a transition rule expressing the edge-matching condition. For this, we need to append to each symbol (i_L, j_L, t_L) a fourth index a_L labeling an edge of the tile $D_{j_L}^{\mathbf{t}_L}(\mathbf{i}_L)$. To construct all possible inadmissible tail sequences, we seek a transition rule of the following sort: given the symbol $((i_L, j_L, t_L, a_L)$, which possible successor symbols $(i_{L+1}, j_{L+1}, t_{L+1}, a_{L+1})$ are compatible with (3.11) and correspond to matching edges in the sense of (3.9) or (3.10)? Ultimately, it should be possible to express the rule without explicit reference to the edge sequences, since the existence of a single sequence of matching edges in the tail is enough to restrict the limit point to a tile boundary. How to accomplish this will become clear in our examples below.

Following the above prescription, one can generate a transition matrix $\mathscr{T}$, with entries 0 and 1, which can be used to test a code σ for admissibility one symbol at a time, at least if the tail consists of a repeated finite symbol string, so that the edge-matching test is guaranteed to terminate. The rows and columns of $\mathscr{T}$ are labeled by the various possible (i,j,t,a); if (i,j,t,a) can be followed by (i',j',t',a') in an inadmissible tail, then the row of $\mathscr{T}$ labeled by (i,j,t,a) has a 1 in the column labeled by (i',j',t',a'); if such a succession of symbols is forbidden, then the entry is a 0. Obviously, if a row (resp. column) consists entirely of 0's, then it and the corresponding column (resp. row) can be deleted, since such a symbol can be banished by initiating the tail sequence at a sufficiently high level.

The transition matrix is easily seen to be equivalent to a graph T with oriented links. The nodes of T are labeled by the possible (i,j,t,a), and if (i,j,t,a) can be followed by (i',j',t',a') in an inadmissible tail, there is a link connecting the corresponding nodes. Any node without an incoming or outgoing link can be deleted, together with its remaining links. The same reasoning allows us to further "reduce" the graph if the latter consists of two subgraphs connected by links which all have the same orientation, as in Fig. 3.2.

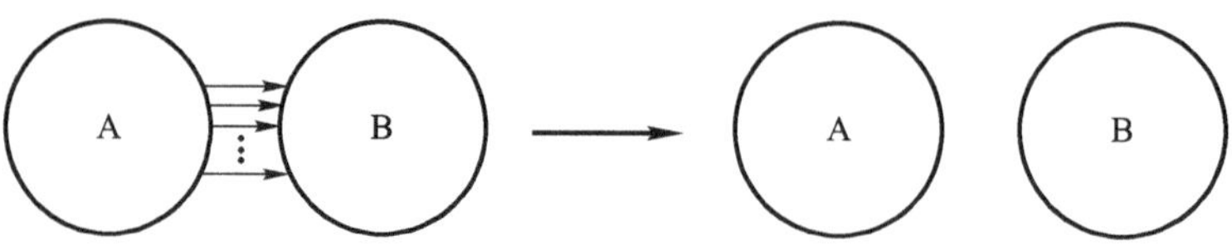

Fig. 3.2 Reduction of a graph.

In that case, the links can be deleted, splitting the graph into disconnected pieces. Repeating the process, deleting isolated nodes along the way, one ends up with a set of non-trivial irreducible subgraphs. For convenience in checking admissibility, one normally converts the transition graph into a table listing the possible successors of symbols which might appear in an inadmissible tail sequence. Let us now see how the above scheme works in two examples. In the first, with quadratic irrational parameter, we derive the rules graphically, while in the second, with cubic irrational parameter, the original transition graph is huge, and it is more feasible to manipulate the corresponding matrix, with computer assistance.

3.3.1 Quadratic example

Our first example is the quadratic model with parameter $\lambda = -\tau = \frac{1}{2}(\sqrt{5}-1)$, the inverse of the golden mean. The catalogue consists of a single triangle, with the do-

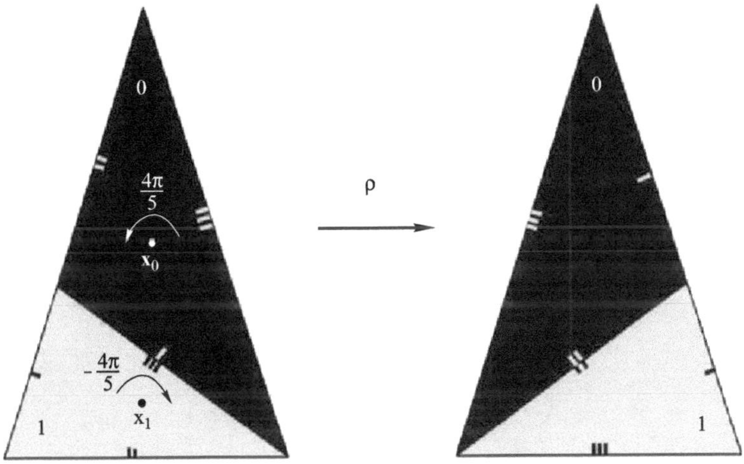

Fig. 3.3 Level-zero return map, with labeled edges. The rotation angles for domains $D_0(0)$ and $D_1(0)$ are $4\pi/5$ and $-4\pi/5$, respectively.

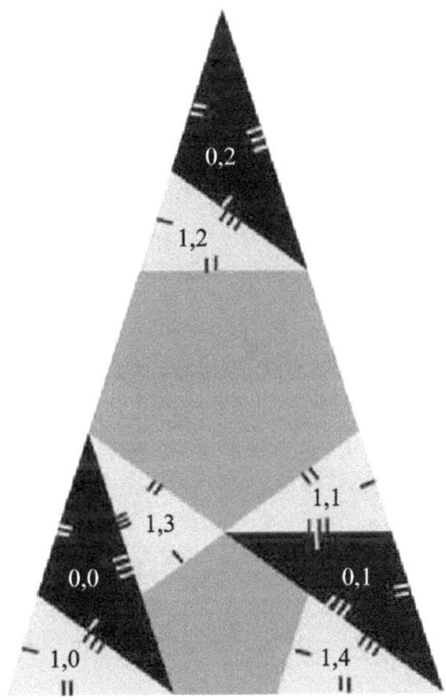

Fig. 3.4 Tiling of level-zero triangle by periodic pentagons and return orbits of level-one subdomains, with labeled edges.

main map shown in Fig. 3.3. To assist the eyes, we have used a coordinate system in which **C** is a true rotation matrix. It is sufficient to consider the edge assignments for levels 0 and 1, shown in Figs. 3.3 and 3.4, since the matching conditions are independent of level. The reader can easily verify visually the edge-matching conditions displayed in the graph of Fig. 3.5. In the latter, the node corresponding to edge a of $D^t_j(0)$ is labeled (jta). If this edge contains an edge a' of the subdomain $D^{t,t'}_{j'}(1)$, the graph has a link directed from (jta) to $(j't'a')$. For example, edge 3 of $D^0_0(0)$ contains edge 3 of $D^{0,2}_0(1)$, edge 1 of $D^{0,1}_1(1)$, and edge 2 of $D^{0,1}_0(1)$. In the graph, this corresponds to links from (013) and (023) to (023), (111), and (012). There is no node (003), since edge 3 of $D^{0,0}_0(1)$ is internal and cannot participate in the true tail of a code sequence.

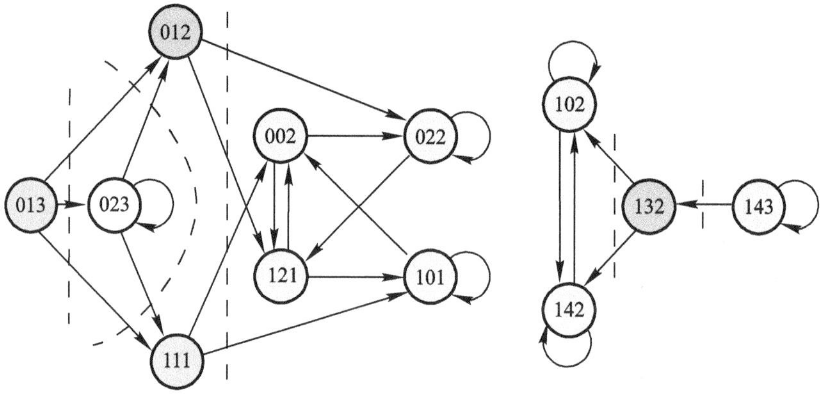

Fig. 3.5 Transition graph for inadmissible codes. Removal of all links crossed by the dashed lines reduces the graph to four connected components. The surviving nodes occupy the white disks.

As explained above, we can reduce the transition graph as follows. Each of the dashed lines splits the graph into two pieces with unidirectional links. Deleting the latter leaves us with 8 irreducible subgraphs, of which 4 are isolated nodes (013),(012),(111),and (132), which can be deleted together with their links. In the figure the four nodes have been displayed as gray disks. The remaining irreducible parts (with nodes represented by white disks) correspond to the following table:

(j_k,t_k)	$(0,2)$	$(0,0),(0,2)'$	$(1,0),(1,2)$	$(1,0)',(1,4)$	$(1,4)'$
(j_{k+1},t_{k+1})	$(0,2)$	$(0,2)',(1,2)$	$(0,0),(1,0)$	$(1,0)',(1,4)$	$(1,4)'$

In each column, any of the symbols above can be followed by any of the symbols below. As a simple application, let us construct the inadmissible codes which are of the form $(j,t)^\infty$. From the table, the condition $(j_k,t_k) = (j_{k+1},t_{k+1})$ selects only $(0,2)^\infty,(1,4)^\infty$, and $(1,0)^\infty$. The admissible period-1 codes are all the remaining ones compatible with the path function $p(0) = (1,0,0), p(1) = (1,0,0,1,1)$, namely $(0,1)^\infty$ and $(1,3)^\infty$.

For period n, what fraction of the codes are inadmissible? We can easily calculate an upper bound for this ratio from the irreducible transition matrices corresponding to the graphs of Fig.3.5, namely

$$(1), \quad (1), \quad \begin{pmatrix} 1 & 1 \\ 1 & 1 \end{pmatrix}, \quad \begin{pmatrix} 0 & 1 & 0 & 1 \\ 0 & 1 & 0 & 1 \\ 1 & 0 & 1 & 0 \\ 1 & 0 & 1 & 0 \end{pmatrix},$$

with respective eigenvalue sets $\{1\}$, $\{1\}$ $\{2,0\}$ $\{2,0,0,0\}$. This gives for the upper bound on the number of inadmissible tail sequences $1+1+2^n+2^n = 2^{n+1}+2$. For $n=1$ this gives 6, which is twice the actual number because of the duplication of unprimed and primed edge labels. The total number of linked sequences is given by Tr $\mathbf{P}^n$, with $\mathbf{P}$ the transition matrix corresponding to the path function. The matrix is 8-dimensional, corresponding to rows and columns labeled by $(j,t) = (0,0),(0,1),(0,2),(1,0),(1,1),(1,2),(1,3),(1,4)$:

$$\mathbf{P} = \begin{pmatrix} 0 & 1 & 1 & 0 & 1 & 1 & 0 & 0 \\ 0 & 1 & 1 & 0 & 1 & 1 & 0 & 0 \\ 0 & 1 & 1 & 0 & 1 & 1 & 0 & 0 \\ 1 & 0 & 0 & 1 & 0 & 0 & 1 & 1 \\ 1 & 0 & 0 & 1 & 0 & 0 & 1 & 1 \\ 1 & 0 & 0 & 1 & 0 & 0 & 1 & 1 \\ 1 & 0 & 0 & 1 & 0 & 0 & 1 & 1 \\ 1 & 0 & 0 & 1 & 0 & 0 & 1 & 1 \end{pmatrix}.$$

The eigenvalues of $\mathbf{P}$ are $\{4,1,0,0,0,0,0,0\}$, and so Tr $\mathbf{P}^n = 4^n + 1$. Thus the fraction of inadmissible sequences is bounded above by $(2^{n+1}+2)/(4^n+1)$, which for large n tends to zero proportional to 2^{-n}.

3.3.2 Models I, II, and III

We have carried out a similar analysis for our three quadratic examples, Models I, II, and III. In each case, a list of edge matches was obtained by iteratively mapping the level-1 atoms along their level-0 return paths. In each case, the list was reduced, with the help of graphs, to a minimal set of succession rules $(j_k,t_k) \to (j_{k+1},t_{k+1})$ for infinitely long tail sequences. The tables allow us to systematically test the admissibility of any given eventually periodic code: if any neighboring code elements in the tail fail to satisfy the listed succession rules, then the code is admissible and we know that the point corresponding to the symbol sequence does not lie on any discontinuity line.

The inadmissibility table computed for Model I is shown in Table 3.1.

Table 3.1 The inadmissibility table for Model I

(j_k,t_k)	$(0,0),(0,1)$	$(2,0),(2,1)$	$(4,4),(4,6)$	$(4,2),(4,4)'$
(j_{k+1},t_{k+1})	$(0,1),(2,1)$	$(0,0),(2,0)$	$(4,2),(4,4)$	$(4,4)',(4,6)$

Note that there are two irreducible sets of code elements, separated by double vertical lines in the table. The corresponding succession matrix has two irreducible blocks, each of the form

$$\mathbf{M} = \begin{pmatrix} 0 & 1 & 0 & 1 \\ 0 & 1 & 0 & 1 \\ 1 & 0 & 1 & 0 \\ 1 & 0 & 1 & 0 \end{pmatrix},$$

with eigenvalues $(2,0,0,0)$. This matrix counts edge matches rather than (j,t) matches, and since we have two distinct edge symbols $(4,4)$ and $(4,4)'$, we have to be on the lookout for duplicate codes. Fortunately, a brief inspection of the possibilities shows that the only periodic code in this category is $(4,4)^\infty$.

Thus, the number of inadmissible codes of period n is

$$I_n = 2\mathrm{Tr}\, M^n - 1 = 2^{n+1} - 1.$$

On the other hand, if $\hat{A}$ is the incidence matrix (see Sect. 2.3) restricted to the open polygonal atoms:

$$\hat{\mathbf{A}} = \begin{pmatrix} 1 & 2 & 4 \\ 1 & 1 & 1 \\ 0 & 1 & 3 \end{pmatrix}$$

with eigenvalues $(4,1,0)$, the total number N_n of period-n codes is given by

$$N_n = \mathrm{Tr}\, \hat{\mathbf{A}}^n = 4^n + 1.$$

We see that the fraction of inadmissible periodic sequences tends to zero as 2^{-n} as n tends to infinity.

For $n = 1$, the formulas give

$$I_1 = 3, \qquad N_1 = 5,$$

which we can check explicitly using Table 3.1 and the Model I path function. The list of codes $(j,t)^\infty$ is the following, with the inadmissible one underlined:

$$\underline{(0,1)}^\infty, \quad \underline{(2,0)}^\infty, \quad (4,2)^\infty, \quad \underline{(4,4)}^\infty, \quad (4,6)^\infty.$$

For $n = 2$, the trace formulas tell us

$$I_2 = 7, \qquad N_2 = 17.$$

Once again, use of Table 3.1 and the path function allow us to identify the admissible and inadmissible (underlined) codes. In addition to the period-1 cases already listed, we have

$$\underline{((0,0)(2,1))}^\infty, \quad \underline{((2,1)(0,0))}^\infty, \quad ((0,0)(2,3))^\infty, \quad ((2,3)(0,0))^\infty,$$
$$((2,2)(4,0))^\infty, \quad ((4,0)(2,2))^\infty, \quad ((4,4)(4,2))^\infty, \quad ((4,2)(4,4))^\infty,$$

$$\underline{((4,2)(4,6))^\infty}, \quad \underline{((4,6)(4,2))^\infty}, \quad ((4,6)(4,4))^\infty, \quad ((4,4)(4,6))^\infty,$$

confirming the results of the matrix calculations.

The other quadratic models can be handled similarly. In particular, the inadmissibility tables for Models II and III, obtained using the edge-matching algorithm, are given in Sect. A.5 of the Appendix. In each case, diagonalization of the succession matrix for inadmissible tails produces eigenvalues which are strictly less than the largest eigenvalue of the restricted incidence matrix. The inadmissible periodic codes thus form a fraction of all codes which tends to zero geometrically as the period increases. It is not difficult to show that the same behavior holds if one restricts oneself to eventually periodic codes.

3.3.3 Cubic example

Our final and most elaborate example is the 4-member catalogue of the $\pi/7$ model. Here the number of domains is too large to calculate easily the matching edges and to visualize accurately the structure of the transition graph. Fortunately, the transition table can still be derived with modest effort, using computer assisted calculation and manipulation of the transition matrices. The computational tools for this are supplied in the electronic supplement to Lowenstein et al. (2004), in the form of Mathematica data structures and functions. Our task is simplified by observing that we may restrict ourselves to two disjoint classes of symbol-sequence tails, namely those with $\mathbf{i} = 6^\infty$ and those with all $i_L \in \{0,1,2,3,4,5\}$. The transition tables for the two classes can be constructed and reduced separately.

Let us work out in detail the nontrivial, irreducible transition tables for the inadmissible codes in the sector $\mathbf{i} = 6^\infty$, this time working with matrices instead of graphs. The first step is to map the domains $D_{j_1}(i_1)$, $i_1 = 6$, around their respective return orbits, recording the matching edges in a table. Each entry of the table corresponds to a particular edge (label a_0) of a particular domain $D^{[3]}_{j_0}$, and lists all edges of the various $D^{t_1}_{j_1}(6)$ which are contained in it. Most pairs (j_0,a_0) contain no such edges, and we omit them from both columns of the table, leaving us with Table 3.2.

Table 3.2 The preliminary inadmissibility table for the $\pi/7$ model, sector $\mathbf{i} = 6^\infty$

(j_0,t_0,a_0)	(j_1,t_1,a_1)
$(0,1,1)$	$(0,1,1)$
$(0,0,2),(0,1,2)$	$(0,1,2),(5,1,2),(5,11,1),(6,1,1),(6,17,1)$
$(5,0,2),(5,1,2),(5,6,2)$	(5,6,2),(6,6,1)
$(5,10,1),(5,11,1)$	$(5,10,1),(6,16,1)$
$(6,t,1), t=0,1,6,16,17$	$(0,0,2),(5,0,2),(6,0,1)$
$(6,0,3)$	$(6,0,3)$
$(8,10,1),(8,22,1)$	$(8,10,1),(8,22,1)$
$(8,22,2),(8,34,2)$	$(8,22,2),(8,34,2)$

The above table corresponds to an 18×18 transition matrix $\mathscr{T}$, which we now proceed to split into irreducible parts. In the matrix $\mathscr{T}^n$, nonzero entries correspond to $(n+1)$-symbol sequences compatible with the recursion rules of inadmissibility. Specifically, if such a sequence begins with α and ends with β, then $\mathscr{T}^n_{\alpha\beta} \neq 0$ for some n. Thus, to reduce the graph, we can examine entries of

$$\mathbf{T} = \mathscr{T} + \varepsilon\mathscr{T} + \varepsilon^2\mathscr{T}^2 + \cdots = \mathscr{T}(1-\varepsilon\mathscr{T})^{-1},$$

with $\varepsilon > 0$ chosen small enough to ensure convergence. If $\mathbf{T}_{\alpha\beta} = 0$, there is no finite inadmissibility-compatible symbol sequence connecting α to β, and we conclude that the corresponding graph splits into subgraphs A and B connected by links directed from B to A, with B containing only those nodes which can initiate a path ending at β.

Explicitly calculating $\mathbf{T}$ with (say) $\varepsilon = .01$, we note there are 2 columns with only a single non-zero entry, and 2 pairs of columns with 2 entries. These produce 4 irreducible sub-matrices, while deleting their 6 rows and columns leaves a 12×12 irreducible matrix, all of whose entries are positive. Stripping off the unnecessary edge labels, we end up with Table 3.3.

Table 3.3 The final inadmissibility table for the $\pi/7$ model, sector $\mathbf{i} = 6^\infty$

(j_0, t_0)	(j_1, t_1)
$(0,0),(0,1)$	$(0,1),(5,1),(5,11),(6,1),(6,17)$
$(5,0),(5,1),(5,6)$	$(5,6),(6,6)$
$(5,10),(5,11)$	$(5,10),(6,16)$
$(6,0),(6,1),(6,6),(6,16),(6,17)$	$(0,0),(5,0),(6,0)$
$(0,1)'$	$(0,1)'$
$(6,0)'$	$(6,0)'$
$(8,10),(8,22)$	$(8,10),(8,22)$
$(8,22)',(8,34)$	$(8,22)',(8,34)$

Here horizontal lines separate the irreducible sub-tables. From the table we can read off the period-1 inadmissible tails, namely

$$(6,0,1)^\infty,(6,5,6)^\infty,(6,5,10)^\infty,(6,6,0)^\infty,(6,8,10)^\infty,(6,8,22)^\infty,(6,8,34)^\infty.$$

From the path constraints, we are left with the following admissible period-1 tails:

$$(6,5,2)^\infty,(6,5,4)^\infty,(6,5,12)^\infty,(6,5,14)^\infty,(6,8,8)^\infty,$$

$$(6,8,12)^\infty,(6,8,14)^\infty,(6,8,30)^\infty,(6,8,32)^\infty,(6,8,36)^\infty.$$

Once again, we can diagonalize the transition matrices and easily calculate an upper bound on the fraction of inadmissible symbol sequences of period n. The ratio is found to decrease for large n as 0.225708^n.

We can proceed in the same way for the tails with all $i_L \in \{0,1,2,3,4,5\}$. The transition matrix, after pruning the rows (resp. columns) with only zeros and the corresponding columns (resp. rows), has dimension 673. The reduction process reveals the simple graphical structure of Fig. 3.6.

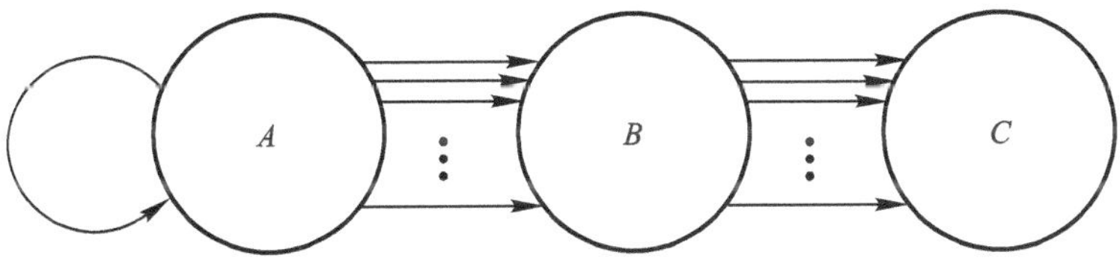

Fig. 3.6 Transition graph for tails with $i_L \in \{0,\ldots,5\}$. The subgraphs A (one node)and C (665 nodes) survive reduction.

where subgraph A contains a single node, B has 7, and C has 665. The single repeating node in A corresponds to tail $(3,2,2)^\infty$. The subgraph B and its links can be deleted, leaving us with a final transition table with $666 = 665 + 1$ rows.

A numerical calculation evaluates the largest eigenvalues of the transition and path matrices to be 8.67058 and 93.3268, respectively, so that the fraction of period-n inadmissible tails tends to zero for large n as 0.0929056^n.

The constraints for period-1 codes $(i,j,t)^\infty$, namely $n(i) = h(i)$ and $p(i,j,t) = j$, restrict us to the following possibilities as shown in Table 3.4.

Table 3.4 The period-1 codes $(i,j,t)^\infty$ of the $\pi/7$ model

i	j	t
1	1	$1-3, 5-7, 14-16$
1	2	$8-13, 21-25, 33-38$
1	7	$0,4,17,29,41,54,66,78,91,103,115$
2	0	$\underline{0}, 2-4, 8, \underline{9}$
2	2	$\underline{6}, 16, 45, 69$
2	3	$7, \underline{17}, 26, 31, 36, 46, 55, 60, 70, 79, 84, 89, 99, \underline{108}, 113$
2	4	$\underline{5}, 15, 25, \underline{30}, 39, 49, 54, 59, 68, 78, 83, 92, 102107, 112$
2	5	$1,11,21,35,45,59,69,88,98$
3	1	$\underline{3}$
3	2	$\underline{1}, \underline{3}$
3	4	$\underline{2}$
3	5	$\underline{0}$
3	7	$\underline{2}$
6	0	$\underline{1}$
6	5	$2, 4, \underline{6}, \underline{10}, 12, 14$
6	6	$\underline{0}$
6	8	$8, \underline{10}, 12, 14, \underline{22}, 30, 32, \underline{34}, 36$

Consulting the admissibility tables obtained in the previous paragraph, we have underlined those entries (i,j,t) which were found to be inadmissible.

3.4 Minimality

For any catalogue member $(D^{[n]}, \rho^{[n]})$, the action of the return map on each $\Sigma^{[n]}(\mathbf{i})$ is not only invariant, but also *minimal* in the sense that, given any $\mathbf{x}, \mathbf{y} \in \Sigma^{[n]}(\mathbf{i})$ and $\varepsilon > 0$, there exists a nonnegative integer m such that $|\rho^m(\mathbf{x}) - \mathbf{y}| < \varepsilon$. This was proved by Lowenstein (2007) making use of the property of the incidence matrices called semi-positivity.

Definition 3.1. (Lowenstein, 2007, Definition 3) A matrix will be called semi-positive if each of its rows has either only zero entries or only positive entries.

Theorem 3.1. (*Lowenstein,* 2007, *Theorem* 1) *Let* $\mathbf{i}$ *be a linked symbol sequence with the property that, for all positive integers K, there exists an integer* $P \geq K$ *such that* $A(i_K, i_{K+1}, \ldots, i_P)$ *is semi-positive. Let* $\rho = \rho^{[n(\mathbf{i})]}$*. Given any* $\mathbf{x}, \mathbf{y} \in \Sigma^{[n]}(\mathbf{i})$ *and* $\varepsilon > 0$*, there exists a non-negative integer m such that* $|\rho^m(\mathbf{x},) - \mathbf{y}| < \varepsilon$.

For the proof, we refer the reader to Lowenstein (2007). There it is also shown that the multi-index incidence matrices $\mathbf{A}(\mathbf{i}_K)$ of the $\pi/7$ model are all semi-positive except for the cases $\mathbf{i}_K = (0,3,3,\ldots,3)$ and $\mathbf{i}_K = (3,3,3,\ldots,3)$. The only possibly nonminimal $\Sigma^{[n]}$ are those for which $\mathbf{i}$ has a tail consisting of all 3's. In fact, however, if one examines the set of codes σ consistent with such sequences $\mathbf{i}$, it turns out that only those ending in $(3,1,3)^\infty$ are standing in the way of minimality. For all others, one can safely replace the matrix elements $A(0)_{0,1}$ and $A(3)_{1,1}$ by zeros and apply Theorem 3.1. But, according to the analysis of the previous section, the code tail $(3,1,3)^\infty$ is *inadmissible*, i.e. it does not actually correspond to points of $\Sigma^{[n]}(\mathbf{i})$. Thus all of the invariant sectors of the $\pi/7$ model are minimal.

References

Afraimovich V., Maass A., and Urias J. (2000) Symbolic dynamics for sticky sets in Hamiltonian systems, *Nonlinearity* **13**, 617–637.

Bratteli O. (1972) Inductive limits of finite-dimensional C* algebras, *Transactions of the American Mathematical Society* **171**, 195–234.

Durand F., Host B., and Skau C. (1999) Substitutional dynamical systems, Bratteli diagrams, and dimension groups, *Ergodic Theory and Dynamical Systems* **19**, 953–993.

Lowenstein J. H. (2007) Aperiodic orbits of piecewise rational rotations of convex polygons with recursive tiling, *Dynamical Systems* **22**, 25–63.

Lowenstein J. H., Kouptsov K. L., and Vivaldi F. (2004) Recursive tiling and geometry of piecewise rotations by $\pi/7$, *Nonlinearity* **17**, 1–25.

Poggiaspalla G. (2006) Self-similarity in piecewise isometric systems, *Dynamical Systems* **21**, 147–189.

Vershik A. M. (1985) A theorem on the Markov periodical approximation in ergodic theory, *Journal of Soviet Mathematics* **28**, 667–674.

Chapter 4
Dimensions and Measures

As we have seen in Chapter 3, the pseudochaotic web (residual set) of aperiodic, discontinuity avoiding orbits in a renormalizable piecewise isometric system, while occupying zero area, nevertheless is far from trivial in its dynamics. In probing more deeply, we are naturally led to ask questions of a probabilistic nature, concerning, for example, what fraction of the web is contained in a particular subset of the plane, or with what probability will an orbit initiated in a particular set return to that set within a designated time? To investigate such questions requires a *measure*, and so we turn our attention to the task of constructing one. Fortunately, *Hausdorff measure*, widely used in the study of Cantor sets and other fractals (Falconer, 1990), turns out to be the most natural choice.

4.1 Hausdorff dimension and Hausdorff measure

The s-dimensional Hausdorff measure of a set D in $\mathbb{R}^n$ is defined (Falconer, 1990) as

$$\mathscr{H}^s(D) = \lim_{\delta \to 0} \mathscr{H}^s_\delta(D), \qquad \mathscr{H}^s_\delta(D) = \inf\left\{\sum_i |U_i|^s : \{U_i\} \text{ is a } \delta\text{-cover of } D\right\}$$

where a δ-cover of D is a finite or countable collection of sets U_i with diameter less than δ:

$$D \subset \bigcup_i U_i, \qquad |U_i| \stackrel{\text{def}}{=} \sup\{|\mathbf{x}-\mathbf{y}| : \mathbf{x},\mathbf{y} \in U_i\} < \delta.$$

For a given D, there exists a unique value of s, the *Hausdorff dimension* $d_H(D)$, such that

$$\mathscr{H}^s(D) = \begin{cases} \infty & \text{if } s < d_H(D), \\ 0 & \text{if } s > d_H(D). \end{cases}$$

If $\mathscr{H}^{d_H(D)}(D)$ is nonzero and finite, we define it to be the *Hausdorff measure of D*.

If Hausdorff measure is well defined (not always the case!), it is clearly invariant with respect to isometries on $\mathbb{R}^n$, and inherits the convenient scaling property of $\mathscr{H}^s$,

$$\mathscr{H}^{d_H}(\omega D) = \omega^{d_H}\mathscr{H}^{d_H}(D).$$

We remind the reader here that "diameter" and "isometry" are to be understood in terms of the $\mathscr{Q}$ metric (2.2). Since the difference is a nonsingular linear transformation, the value of d_H and the existence of Hausdorff measure are the same as for the Euclidean metric, but obviously the measure of a particular set does depend on the choice of metric.

4.2 Construction of the measure

We now restrict our attention to a renormalizable piecewise isometry of the square, with a hierarchy of dressed domains $\Delta(i_1,\ldots,i_L) = (D(i_1,\ldots,i_L),\rho(i_1,\ldots,i_L))$ rooted in a catalogue $\{\Delta^{[n]} : n = 0,\ldots,N-1\}$. The return map structure of the recursive tiling hierarchy is characterized by an incidence matrix $A_{j,k}(i)$ specifying the number of times the return orbit of $D_k(i)$ visits the atom $D_j^{[n(i)]}$, where $n(i)$ is defined by $D(i) \subset D^{[n(i)]}$. Scaling within the hierarchy is governed by the $\mathscr{Q}$-metric diameter ratios $\omega(i) = |D(i)|/|D^{[h(i)]}|$, where $h(i)$ is the catalogue member similar to $D(i)$.

We would like to define a measure μ on the square which vanishes on the periodic cells and boundary lines of tiles, but coincides with Hausdorff measure on the residual set (hence on the closure of the latter, the exceptional set). We proceed to construct such a measure as in Lowenstein et al. (2004).

We start with the set $\mathscr{S}$ of tiles and cells of arbitrary level, their edges and vertices, and the empty set. Thanks to the recursive tiling structure, the set is a *semiring*, since it is closed under intersection and, for all $A,B \in \mathscr{S}$ such that $B \subset A$, the set $A \setminus B$ is a finite or countable union of disjoint elements of $\mathscr{S}$. We define μ by setting $\mu(A) = 0$ if A is a cell, edge, vertex, or empty, and assign a positive number $\psi_{n,j}$ (to be determined below) to each level-0 atom $D_j^{[n]}(i)$, extending the definition to arbitrary tiles by assuming isometric invariance and the scaling relation

$$\mu(D_j(i_1,\ldots,i_L)) = \omega(i_1,\ldots,i_L)^s\mu(h(D_j(i_1,\ldots,i_L))), \tag{4.1}$$

where the "dimension" s is another parameter to be determined.

To complete the definition of μ, we apply the additivity equations expressing the root level of the recursive tiling property,

$$\sum_{\substack{i,k\\ n(i)=n}} A_{j,k}(i)\mu(D_k(i)) = \mu(D_j^{[n]}), \qquad n = 1,\ldots,N, \qquad j = 0,\ldots,J(n)-1. \tag{4.2}$$

Inserting the unknown $\psi_{n,j}$ and the scaling relation (4.1), Equation (4.2) assumes the form of a matrix eigenvalue equation,

$$\sum_{(n',j')} M_{(n,j),(n',j')}(s)\psi_{n',j'} = \psi_{n,j}, \tag{4.3}$$

where (treating the pair (n, j) as a single matrix index),

$$M_{(n,j),(n',j')}(s) = \sum_i{}' \delta_{n,n(i)}\delta_{n',h(i)}A_{j,j'}(i)\omega(i)^s. \tag{4.4}$$

If the determinental condition for a nontrivial solution,

$$\det(\mathbf{1} - \mathbf{M}(s)) = 0,$$

has a solution $s = d_H$, then the $\psi_{n,j}$ can be calculated from the linear system (4.2) up to a common normalization factor.

It is now straightforward, using the recursive tiling property to establish σ-additivity, to prove that if all the $\psi_{n,j}$ are strictly positive, μ is a genuine probability measure. The final step, to identify the restriction of μ to the residual set Σ with Hausdorff measure (up to normalization), is more tedious, and we refer the reader to Lowenstein et al. (2004) , Theorem 4, for the detailed proof. The main strategy is to show that for an arbitrary tile D, there exist positive numbers κ_1 and κ_2 such that the measure $\mathscr{H}^{d_H}(D \cap \Sigma)$ satisfies the inequalities

$$\kappa_1\mu(D) \leq \mathscr{H}^s(D \cap \Sigma) \leq \kappa_2\mu(D),$$

and so is nonzero and finite. Since Hausdorff measure must also satisfy the additivity condition (4.2), and the solution is unique up to normalization, we have the desired result.

We see that the problem of constructing a suitable measure for the residual set reduces to an eigenvalue problem for the *transfer matrix* $\mathbf{M}(s)$ with components given by (4.4). Let us summarize the result as a theorem (after Lowenstein et al., 2004, Theorem 4 and Corollary 5):

Theorem 4.1. *Let* $\mathbf{M}(s)$ *be the transfer matrix for a catalogue with recursive tiling. Suppose that, for some* $s = d_H, 1 < d_H \leq 2$, *there exists an* $\mathbf{M}(d_H)$*-invariant vector* ψ *whose components are positive. Then a finite, nonzero Hausdorff measure* $\mathscr{H}_H^d$ *can be introduced on the sigma-ring generated by the restrictions of all tiles to the residual set. On the lowest-level restricted tiles, the Hausdorff measure coincides with the elements of* ψ, *up to overall normalization. The Hausdorff dimension* d_H *is a solution of the transcendental equation* $\det(\mathbf{1} - \mathbf{M}(s)) = 0$.

4.3 Simplification for quadratic irrational λ

For the models with quadratic irrational parameter which we have studied, the calculation of Hausdorff dimension and Hausdorff measure simplifies considerably.

For most of these models, the recursive tiling structure reduces to a simple scaling sequence, with a single scale factor ω and a single incidence matrix $\mathbf{A}$. The transfer matrix then takes the form

$$\mathbf{M}(\mathbf{s}) = \omega^s \mathbf{A},$$

so that the eigenvalue problem of Theorem 4.1 reduces to

$$\mathbf{A} \cdot \psi = \omega^{-s} \psi.$$

If diagonalization of $\mathbf{A}$ produces an eigenvalue a in the interval $(1,2]$, with corresponding (non-normalized) eigenvector $\psi(a)$, we obtain a Hausdorff dimension

$$d_H = \frac{\log a}{|\log \omega|}$$

and a measure determined, up to an overall normalization factor c, by

$$\mu(D_j(0)) = c\psi(a)_j.$$

For a probability measure on the square, we have

$$c^{-1} = \sum_{k=0}^{J-1} T_k(0)\psi(a)_k,$$

where $T_k(0)$ is the level-0 return time for the kth atom of $D(0)$.

To illustrate the method, consider Model I from Sec. 2.3 with parameter $\lambda = (1+\sqrt{5})/2$. Here the scale factor is $\omega = \lambda^{-2}$, and the reduced incidence matrix, restricted to the polygonal domains $D_j(0)$, $j = 0,2,4$,

$$\mathbf{A} = \begin{pmatrix} 1 & 2 & 4 \\ 1 & 1 & 1 \\ 0 & 1 & 3 \end{pmatrix}$$

has eigenvalues $4,1,0$. Only the first leads to a dimension in the correct range, namely

$$d_H = \frac{\log 2}{\log \lambda}.$$

The corresponding eigenvector of $\mathbf{A}$ is, up to normalization

$$\psi = (2,1,1).$$

Since the level-0 return times are (see Sect. A.1.2 of the Appendix) 6, 10, and 18, we obtain

$$\{\mu(D_j(0)) : j = 0,2,4\} = \left\{\frac{1}{20}, \frac{1}{40}, \frac{1}{40}\right\}.$$

4.4 A complicated example: Model II

The only quadratic example which is slightly more complicated than the preceding one is Model II from Sec. 2.3, $\lambda = \sqrt{3}$ and $\tau = -\sqrt{3}/3$, where there are two non-overlapping pseudochaotic webs, each with its own incidence matrix. We construct two separate measures, each coinciding with Hausdorff measure on one of the webs and vanishing on the other. We begin with hierarchy A, with scaling limit (τ, τ). From (A.1), the reduced incidence matrix is

$$\mathbf{A} = \begin{pmatrix} 0 & 0 & 0 & 0 & 0 & 4 & 2 & 5 & 1 & 0 & 0 & 0 \\ 0 & 0 & 0 & 0 & 0 & 24 & 12 & 25 & 7 & 0 & 0 & 0 \\ 0 & 0 & 0 & 0 & 0 & 5 & 2 & 4 & 1 & 0 & 0 & 0 \\ 0 & 1 & 2 & 1 & 1 & 1 & 1 & 1 & 1 & 1 & 1 & 1 \\ 0 & 0 & 0 & 1 & 2 & 1 & 1 & 1 & 1 & 1 & 1 & 0 \\ 0 & 0 & 0 & 0 & 0 & 0 & 0 & 0 & 0 & 0 & 0 & 0 \\ 0 & 0 & 0 & 0 & 0 & 0 & 0 & 0 & 0 & 0 & 0 & 0 \\ 0 & 0 & 0 & 0 & 0 & 0 & 0 & 0 & 0 & 0 & 0 & 0 \\ 0 & 0 & 0 & 0 & 0 & 0 & 0 & 0 & 0 & 0 & 0 & 0 \\ 0 & 0 & 0 & 0 & 0 & 34 & 19 & 34 & 13 & 1 & 0 & 0 \\ 0 & 0 & 0 & 0 & 1 & 1 & 1 & 1 & 1 & 1 & 2 & 1 \\ 1 & 1 & 1 & 1 & 1 & 1 & 1 & 1 & 1 & 1 & 1 & 1 \end{pmatrix}.$$

Diagonalizing this matrix, we find that the eigenvalues are 4, 1 (3 times), and 0 (8 times). The largest of these yields a Hausdorff dimension

$$d_H = -\frac{\log 4}{\log \omega} = \frac{\log 4}{\log(2+\sqrt{3})} = 1.05\ldots$$

and a measure specified on the atoms by

$$\psi = (0,0,0,1,1,0,0,0,0,0,1,1),$$

with normalization factor c^{-1},

$$c = T_7(0) + T_{10}(0) + T_{21}(0) + T_{22}(0) = 161280.$$

Note that only the atoms $D_j(0)$, $j = 7, 10, 21, 22$ contribute to the measure. This can be understood in terms of the pattern of zeros in $\mathbf{A}$. By a permutation of the labels of the atoms, the matrix can be transformed into the block form

$$\mathbf{A} \rightarrow \begin{pmatrix} \mathbf{B} & \mathbf{C} \\ \mathbf{O} & \mathbf{D} \end{pmatrix},$$

where **B** is the 4×4 submatrix corresponding to $j = 7, 10, 21, 22$, and **O** has only zero elements. Thus the diagonalization factorizes, with the eigenvalue 4 arising from **B** alone.

For hierarchy B, with scaling limit $(41 + 72\tau, 41 + 72\tau)$, the reduced incidence matrix, corresponding to the atoms $D_j,\ j = 0, 3, 5, 6, 8, 10$, is

$$\mathbf{A} = \begin{pmatrix} 0 & 0 & 0 & 1 & 0 & 0 \\ 0 & 2 & 2 & 2 & 0 & 0 \\ 0 & 1 & 1 & 1 & 1 & 2 \\ 1 & 1 & 1 & 1 & 1 & 1 \\ 0 & 1 & 2 & 3 & 0 & 0 \\ 0 & 0 & 1 & 0 & 0 & 0 \end{pmatrix},$$

with largest eigenvalue 5 and corresponding eigenvector

$$\psi = (3, 20, 15, 15, 19, 3).$$

The Hausdorff dimension of this portion of the residual set is thus

$$d_H = -\frac{\log 5}{\log \omega} = \frac{\log 5}{\log(2 + \sqrt{3})} = 1.22\ldots.$$

The Hausdorff measure on the six atoms is specified by $c^{-1}\psi$, with

$$c = 3\,T_0(0) + 20\,T_3(0) + 15\,T_5(0) + 15\,T_6(0) + 19\,T_8(0) + 3\,T_{10}(0) = 5298000.$$

4.5 Discontinuity set in Model III

In Chapter 2, we discussed in detail the dynamics on the discontinuity set of Model III, a piecewise isometry on $(\tau, -\tau]^2$, $\tau = -\sqrt{2}/2$, with scale factor $\omega = 3 + 4\tau$. The complement of the periodic orbits on the 'base line' $\mathscr{B}$ (the right-hand edge of Ω) was found to be a Cantor set $\mathscr{C}$ with apparent Hausdorff dimension $d_H = \log 5/\log \omega$. Let us now verify that result using the transfer matrix method, simultaneously obtaining the corresponding measure on the base line.

The scaling on the discontinuity set was found to involve the 2-dimensional domains $D_j(n)$, $j = 1, 3, 5, 7$, plus boundary points, which we can safely ignore in the measure calculation. For both even and odd levels, the reduced incidence matrix takes the form

$$\mathbf{A} = \begin{pmatrix} 1 & 2 & 3 & 1 \\ 0 & 1 & 1 & 1 \\ 2 & 2 & 2 & 1 \\ 1 & 1 & 0 & 1 \end{pmatrix}$$

with largest eigenvalue 5 and corresponding eigenvector

$$\psi = (3,1,3,1).$$

According to Sect. A.3.7, the base line is partitioned, up to discrete points, into 3 copies of $D_j(-1)$, $j = 0,2,6$, and a single copy of $D_5(-1)$. Together, these are partitioned into a periodic segment (coinciding with $D_5(-1)$) plus 6 copies of $D_1(0)$, 9 copies of $D_3(0)$, 9 copies of $D_5(0)$, and 6 copies of $D_7(0)$. This gives us a normalized measure with values $c^{-1}\psi$ on the level-zero atoms, with normalization factor

$$c = 3\times 6 + 1\times 9 + 3\times 9 + 1\times 6 = 60.$$

4.6 Multifractal residual set of the $\pi/7$ model

As one might suspect, it is the $\pi/7$ model, with its more elaborate renormalization tree, which provides a more challenging application of the transfer matrix construction of Hausdorff measure on the residual set. As we have seen in Chapter 2, the latter decomposes into an uncountable infinity of disjoint invariant sets, and the complex scaling behavior is best described by a multifractal formalism with a continuous spectrum of dimensions rather than the one or two Hausdorff dimensions which we encounter in the quadratic models. Nevertheless, it is instructive, to see how our transfer matrix methods lead (Lowenstein et al., 2004) to a fairly simple transcendental equation for the overall Hausdorff dimension.

We begin by determining the transfer matrix with the help of the tables in Sect. A.4 of the Appendix, and Figs. 2.12, 2.13 and 2.16. We note that the catalogue member $\Delta^{[0]}$ appears only at the lowest level of the tiling of Ω, and so will not contribute to the calculation of d_H. Its incidence matrices $\mathbf{A}(i)$, $i = 9,\ldots,14$, will not be needed, except at the end for imposing the normalization of the measure. From (4.4), we have the transfer matrix, in block form,

$$\begin{pmatrix} \omega(1)^s\mathbf{A}(1)+\omega(2)^s\mathbf{A}(2) & \omega(0)^s\mathbf{A}(0) & \mathbf{0} \\ \omega(4)^s\mathbf{A}(4)+\omega(5)^s\mathbf{A}(5) & \omega(3)^s\mathbf{A}(3) & \mathbf{0} \\ \omega(7)^s\mathbf{A}(7)+\omega(8)^s\mathbf{A}(8) & \mathbf{0} & \omega(6)^s\mathbf{A}(6) \end{pmatrix}$$

The zeros in the third multi-column correspond to the fact that in the renormalization tree, there are no dressed domains similar to $\Delta^{[3]}$ appearing as children of domains similar to $\Delta^{[1]}$ or $\Delta^{[2]}$. As a consequence, the diagonalization of $\mathbf{M}(s)$ factorizes. In fact (as we shall see below), it is only the upper left, 23×23 submatrix which is relevant.

The problem can be further simplified by expressing the 6 scale factors $\omega(i)$ in terms of the fundamental units η_1,η_2 from (2.14):

$$\omega(0) = \eta_1\eta_2^3, \quad \omega(1) = \omega(2) = \eta_1^2\eta_2^2, \quad \omega(3) = \eta_1^{-1}\eta_2^3, \quad \omega(4) = \omega(5) = \eta_2^2.$$

The determinantal equation $D(s) = \det(\mathbf{1} - \mathbf{M}(s)) = 0$ then becomes

$$F(\eta_1^s, \eta_2^s) = 0,$$

where F is the polynomial

$$\begin{aligned} F(x,y) = {} & 1 - 6y^2 - 37x^2y^2 - 51xy^3 + 9y^4 + 92x^2y^4 - 29x^4y^4 + 71xy^5 \\ & -69x^3y^5 - 4y^6 - 66x^2y^6 + 23x^4y^6 + 9x^6y^6 - 20xy^7 - 48x^3y^7 \\ & -50x^5y^7 + 25x^2y^8 + 19x^4y^8 - 9x^6y^8 + 5x^3y^9 - 6x^5y^9 - 6x^4y^{10}, \end{aligned}$$

which is irreducible over the rationals. To show the existence of a root between $s = 1$ and $s = 2$, we can evaluate $D(1)$ and $D(2)$ as polynomials in λ, using the cubic equation for λ to reduce the polynomial to second degree. One finds

$$D(1) = 20345 + 3096\lambda - 7985\lambda^2 < 0, \tag{4.5}$$

$$D(2) = -7462700 - 1803627\lambda + 3299289\lambda^2 > 0. \tag{4.6}$$

Moreover, by bounding the derivative of $D(s)$ in the interval $(1,2)$, one can prove that the function has a simple zero $s = d_H$ there. It is then a trivial matter, using numerical root-finding, to evaluate d_H to as many digits of precision as desired. With 20 significant digits, the result is

$$d_H = 1.6522336518816620573... \tag{4.7}$$

with the corresponding array of un-normalized Hausdorff measures of the 23 atoms of $\Delta^{[1]}$ and $\Delta^{[2]}$ (displayed with 6-digit precision):

$$\begin{aligned} \psi = {} & (12.2300, 6.22136, 5.56962, 3.20561, 3.20561, 2.97613, 1.30659, 1.22344, \\ & 0.428586, 0.771474, 0.771474, 8.56381, 28.5357, 29.2811, 16.4917, \\ & 20.8323, 9.14583, 5.94682, 4.40014, 2.61159, 3.27812, 2.67410, 1) \end{aligned}$$

It is now a straightforward but tedious exercise to construct the normalized measure on the atoms of all catalogue members, but we will not continue along that path.

4.7 Asymptotic factorization

In a renormalizable model with recursive tiling, the geometric scaling is exact. Each link of the recursion tree is labeled by an index $i \in \{0, 1, \ldots, M-1\}$, where M is the total number of parent-child relationships among the N members of the catalogue. Corresponding to each i is a unique scale factor $\omega(i)$. The diameter of any tile $D_j^{t_0,\ldots,t_L}(i_0,\ldots,i_L)$, relative to that of $D^{[h(i_L)]}$, is given by the cumulative scale factor along the path $i_0,\ldots,i_L$, namely

$$\omega(\mathbf{i}_L) = \prod_{k=0}^{L} \omega(i_k).$$

The independence of $\mathbf{t}_L$ is the result of isometric invariance, provided that we take care to measure diameters using the $\mathscr{Q}$ metric.

For the calculation of measures and dimensions, the geometric scaling is, of course, only one of the two essential ingredients, the other being *temporal recursion.* Here we avoid using the word *scaling*, since in general there will not exist individual factors associated with links of the tree. Rather we are dealing with $J(n(i)) \times J(h(i))$ rectangular matrices, namely the incidence matrices $\mathbf{A}(i)$, which allow us to calculate the level-$(L+1)$ return times as linear combinations of those of level L:

$$T_j(\mathbf{i}_L) = \sum_{k=0}^{J(n(i_L))} T_k(\mathbf{i}_{L-1}) A_{k,j}(i_L),$$

or, in matrix notation,

$$\mathbf{T}(\mathbf{i}_L) = \mathbf{T}(\mathbf{i}_{L-1}) \cdot \mathbf{A}(i_L).$$

The cumulative incidence matrix along the path $i_0, \ldots, i_L$ is just the product, in left-to-right order, of the matrices $\mathbf{A}(i)$,

$$\mathbf{A}(\mathbf{i}_L) = \mathbf{A}(i_0)\mathbf{A}(i_1)\cdots\mathbf{A}(i_L). \tag{4.8}$$

Summing over the row index, one obtains a list of the total $\rho^{[n(i_0)]}$ return times of the tiles $D_j(\mathbf{i}_L)$, $j = 0, \ldots, J(h(i_L))$. Summing over both indices, one gets the *norm* $||\boldsymbol{A}(\mathbf{i}_L)||$, the total number of tiles in the return orbit of the base domain $\Delta(\mathbf{i}_L)$.

While the matrix character of the temporal recursion relations is a complication in the calculation of dimensions and measures, it is not necessarily a troubling one. This is evident in all of the quadratic models we have studied, for reasons which are not difficult to understand. In those models, the i dependence is trivial, so that the matrix product (4.8) become a simple matrix power $\mathbf{A}^{L+1}$. Moreover, the incidence matrix, in all of our quadratic examples, is a square matrix with a single largest eigenvalue $a > 0$, and so, when applied n times on a vector $\boldsymbol{\psi}$, gives, for asymptotically large n,

$$\mathbf{A}^n \cdot \boldsymbol{\psi} \sim a^n \boldsymbol{\psi}_0,$$

where $\boldsymbol{\psi}_0$ is a vector proportional to the eigenvector associated with a. Another way of looking at the situation is the following. Suppose we make a Jordan decomposition of $\mathbf{A}$,

$$\mathbf{A} = \mathbf{S} \cdot \mathbf{A}_0 \cdot \mathbf{S}^{-1}, \tag{4.9}$$

where the invertible matrix $\mathbf{S}$ can be chosen so that $\mathbf{A}_0$ has a triangular form with vanishing elements below the main diagonal and the largest eigenvalue placed at the lower right-hand corner of the matrix. For asymptotically large n, the leading contribution to $\mathbf{A}^n$ in Jordan form will be

$$a^n\mathbf{S}\cdot\begin{pmatrix}0&0&\cdot&\cdot&\cdot&0\\0&0&\cdot&\cdot&\cdot&0\\\cdot&&\cdot&&&\cdot\\\cdot&&&\cdot&&\cdot\\\cdot&&&&0&0\\0&0&\cdot&\cdot&\cdot&1\end{pmatrix}\cdot\mathbf{S}^{-1}=a^n\,\mathbf{u}\otimes\mathbf{v},$$

where

$$(\mathbf{u}\otimes\mathbf{v})_{k,j}\stackrel{\text{def}}{=}u_kv_j,\qquad \mathbf{u}=(S_{0,J},\ldots,S_{J,J}),\qquad \mathbf{v}=(S^{-1}_{J,0},\ldots,S^{-1}_{J,J}).$$

Note $\mathbf{u}\cdot\mathbf{v}$ is the J,J element of the identity matrix, hence is equal to unity. Moreover, $\mathbf{u}$ is an eigenvector of the leading part of $\mathbf{A}^n$ with eigenvalue a^n , since

$$(\mathbf{u}\otimes\mathbf{v})\cdot\mathbf{u}=(\mathbf{v}\cdot\mathbf{u})\mathbf{u}=\mathbf{u}.$$

It is obviously too much to expect that arbitrary products (4.8), constrained only by the linkage condition $J(h(i_k))=J(n(i_{k+1}))$, should have the asymptotic simplicity of simple powers. Nevertheless, it has been possible to extract a factorization theorem (Lowenstein, 2007) which states that, if certain technical inequalities are satisfied by $\mathbf{A}(i)$, then the product matrix $\mathbf{A}(\mathbf{i}_L)$ factorizes for asymptotically large L, with an error which decreases as δ^L, with $\delta<1$. The theorem does not imply a limiting matrix for $\mathbf{A}(\mathbf{i}_L)/||\mathbf{A}(\mathbf{i}_L)||$, but rather allows the second factor in the $\mathbf{u}(\mathbf{i}_L)\otimes\mathbf{v}(\mathbf{i}_L)$ tensor product to fluctuate while the first factor approaches a limiting value, $\mathbf{u}_\infty(\mathbf{i})$. This will be enough to construct a unique invariant measure supported on the closure of $\Sigma(\mathbf{i})$, as well as to apply transfer matrix techniques to investigate the multifractal dimensions and measures on the full residual set.

We begin by introducing some important definitions from Lowenstein (2007) to be used in the factorization theorem. Since the dynamical origin of the matrices is irrelevant here, we will use the notation $\mathbf{B}(i)$ instead of $\mathbf{A}(i)$.

Definition 4.1. Let $\mathbf{B}(i), i=0,1,\ldots,M-1$, be a finite set of $p(i)\times q(i)$ matrices with non-negative coefficients. Let $\mathbf{B}(\mathbf{i}_N)\stackrel{\text{def}}{=}\mathbf{B}(i_1)\cdot\mathbf{B}(i_2)\cdots\mathbf{B}(i_N)$. Define

$$||\mathbf{B}(\mathbf{i}_N)||=\sum_{k,j}B(\mathbf{i}_N)^j_k,\ u(\mathbf{i}_N)_k=\sum_j B(\mathbf{i}_N)^j_k/||\mathbf{B}(\mathbf{i}_N)||,\ v(\mathbf{i}_N)^j=\sum_k B(\mathbf{i}_N)^j_k/||\mathbf{B}(\mathbf{i}_N)||,$$

$$\mathbf{w}(\mathbf{i}_N)=\frac{\mathbf{B}(\mathbf{i}_N)}{||\mathbf{B}(\mathbf{i}_N)||}-\mathbf{u}(\mathbf{i}_N)\otimes\mathbf{v}(\mathbf{i}_N)^T,$$

$$r(\mathbf{i}_N)=\max_{j,k}|w(\mathbf{i}_N)^j_k|,\qquad \eta=\max\{r(i):i=0,\ldots,M-1\},\tag{4.10}$$

$$\Delta\mathbf{u}(\mathbf{i}_N)=\mathbf{u}(\mathbf{i}_N)-\mathbf{u}(\mathbf{i}_{N-1}),$$

$$\Delta\mathbf{v}(\mathbf{i}_N)=\mathbf{v}(\mathbf{i}_N)-\mathbf{v}(i_N).$$

The factorization theorem will take as its hypothesis certain properties of the individual matrices $\mathbf{B}(i)$. The most important of these is $r(i)$, which is a measure of how well the $\mathbf{B}(i)$ are already factorizable. If this quantity is sufficiently small, and in addition the inner products $\mathbf{v}(i)\cdot\mathbf{u}(j)$ are not too small, it is likely that the family of matrices $\mathbf{B}(i)$ will be *well conditioned* in the sense of the following definition.

Definition 4.2. A family of matrices $B(i)$ as in Definition 4.1 will be called well conditioned if the following relations are satisfied:
(i) *For* $i,j=0,1,\dots,M-1$, *such that* $p(j)=q(i)$,

$$\varepsilon_{\pm}(i,j) \stackrel{\text{def}}{=} \frac{1}{2}\left(\mathbf{v}(i)\cdot\mathbf{u}(j) \pm \sqrt{(\mathbf{v}(i)\cdot\mathbf{u}(j))^2 - 4r(j)}\right)$$

is real, and moreover,

$$\varepsilon \stackrel{\text{def}}{=} \max_{i,j} \varepsilon_-(i,j) < \min_{i,j} \varepsilon_+(i,j).$$

(ii) *The following inequalities hold:*

$$c>0, \quad V>\varepsilon, \quad \delta<1,$$

where

$$\begin{aligned}
c &= \max\{(\mathbf{v}(i)\cdot\mathbf{u}(j)-\varepsilon)^{-1} : i,j=0,\dots,M-1, p(j)=q(i)\},\\
V &= \min\{v(i)^k : i=0,\dots,M-1,\ k=0,\dots,q(i)\},\\
\delta &= \max\{p(j)^2 r(j)(\mathbf{v}(i)\cdot\mathbf{u}(j)-\varepsilon)^{-2} : i,j=0,\dots,M-1, p(j)=q(i)\}.
\end{aligned}$$

Let us now state and prove the factorization theorem of Lowenstein (2007).

Theorem 4.2. *Let* $\{\mathbf{B}(i) : i=0,\dots,M-1\}$ *be a well conditioned family of non-negative* $p(i)\times q(i)$ *matrices, and let* $\mathbf{i}=(i_1,i_2,\dots)$ *be an infinite sequence of integers* i_k *with* $0\le i_k<M$ *and* $p(i_{k+1})=q(i_k)$ *for all k. Then, in the notation of Definitions* 4.1 *and* 4.2,

$$||\mathbf{B}(\mathbf{i}_N)|| = ||\mathbf{B}(\mathbf{i}_{N-1})|| \, ||\mathbf{B}(i_N)|| \mathbf{v}(\mathbf{i}_{N-1})\cdot\mathbf{u}(i_{N-1}), \qquad N>1,$$

and, for $k=0,\dots,p(i_1)-1, j=0,\dots,q(i_N)-1$,

$$\begin{aligned}
r(\mathbf{i}_N) &\le \eta\delta^{N-1}. &(4.11)\\
|\Delta u(\mathbf{i}_N)_k| &\le c\eta\delta^{N-2}, &(4.12)\\
|\Delta v(\mathbf{i}_N)^j| &\le \varepsilon, &(4.13)
\end{aligned}$$

Moreover,

$$\frac{\mathbf{B}(\mathbf{i}_N)}{||\mathbf{B}(\mathbf{i}_N)||} = \mathbf{u}(\mathbf{i}_N)\otimes\mathbf{v}(\mathbf{i}_N) + O(\delta^N),$$

and there exists a vector $\mathbf{u}_\infty(\mathbf{i})$ *such that*

$$|\mathbf{u}(\mathbf{i}_N)-\mathbf{u}_\infty(\mathbf{i})| = O(\delta^N).$$

Proof. We begin by deriving some useful estimates. From the definitions, the components of $\mathbf{u}(\mathbf{i}_N)$ sum up to unity, as do those of $\mathbf{v}(\mathbf{i}_N)$. Hence

$$\sum_k w(\mathbf{i}_N)_k^j = \sum_j w(\mathbf{i}_N)_k^j = 0.$$

Moreover, if a represents the arguments $i_1,\ldots,i_K$, and b denotes $i_{K+1},\ldots,i_N$, then the following identities follow from substitution into the product formula, $\mathbf{B}(\mathbf{i}_N) = \mathbf{B}(a)\cdot\mathbf{B}(b)$, and the majorization of the components of $w(a)$ and $w(b)$ by $r(a)$ and $r(b)$:

$$||\mathbf{B}(\mathbf{i}_N)|| = ||\mathbf{B}(a,b)|| = ||\mathbf{B}(a)||||\mathbf{B}(b)||\mathbf{v}(a)\cdot\mathbf{u}(b), \tag{4.14}$$

$$|u(a,b)_k - u(a)_k| = \left|\frac{(\mathbf{w}(a)\cdot\mathbf{u}(b))_k}{\mathbf{v}(a)\cdot\mathbf{u}(b)}\right| \le \frac{r(a)}{\mathbf{v}(a)\cdot\mathbf{u}(b)}, \tag{4.15}$$

$$|v(a,b)^j - v(b)^j| = \left|\frac{(\mathbf{v}(a)\cdot\mathbf{w}(b))^j}{\mathbf{v}(a)\cdot\mathbf{u}(b)}\right| \le \frac{r(b)}{\mathbf{v}(a)\cdot\mathbf{u}(b)}. \tag{4.16}$$

$$\begin{aligned}|w(\mathbf{i}_N)_k^j| &= (\mathbf{v}(a)\cdot\mathbf{u}(b))^{-2}\Big|\sum_{k_1,k_2,j_1,j_2} w(a)_k^{j_1} v(a)^{j_2} w(b)_{k_1}^j u(b)_{k_2}(\delta_{k_1}^{j_1}\delta_{k_2}^{j_2} - \delta_{k_1}^{j_2}\delta_{k_2}^{j_1})\Big| \\ &\le (\mathbf{v}(a)\cdot\mathbf{u}(b))^{-2}\sum_{k_1,j_1}|w(a)_k^{j_1}||w(b)_{k_1}^j| \\ &\le q(a)p(b)r(a)r(b)(\mathbf{v}(a)\cdot\mathbf{u}(b))^{-2}.\end{aligned} \tag{4.17}$$

Equipped with the above estimates, we now prove (4.11), (4.12), and (4.13).

The bound (4.11) clearly holds for $N=1$, by definition. Moreover, if we assume $r(\mathbf{i}_{N-1}) \le \eta\delta^{N-2}$, we can use (4.17) and the definition of δ to obtain, for any j,k,

$$|w(\mathbf{i}_N)_k^j| \le r(\mathbf{i}_{N-1})\delta \le \eta\delta^{N-1},$$

and thus,

$$r(\mathbf{i}_N) \le \eta\delta^{N-1}.$$

This establishes (4.11) by induction.

To get (4.12) for arbitrary k and $N>1$, we apply (4.15), (4.11) and the definition of c:

$$|\Delta u(\mathbf{i}_N)_k| \le \frac{\eta\delta^{N-2}}{\mathbf{v}(i_{N-1})\cdot\mathbf{u}(i_N) - \varepsilon} \le c\eta\delta^{N-2}.$$

For the third estimate we again resort to mathematical induction. For $N=2$ and arbitrary j, we have

$$|\Delta v(i_1,i_2)^j| \le \frac{r(i_2)}{\mathbf{v}(i_1)\cdot\mathbf{u}(i_2)} \le \frac{r(i_2)}{\mathbf{v}(i_1)\cdot\mathbf{u}(i_2) - \varepsilon} < \varepsilon,$$

where the final inequality is a consequence of the definition of ε and its assumed upper bound, which imply, for all j,k such that $p(k) = q(j)$,

$$\varepsilon^2 - \mathbf{v}(j) \cdot \mathbf{u}(k)\,\varepsilon + r(k) < 0.$$

Since $\varepsilon < v(j) \cdot u(k)$, we have

$$\frac{r(k)}{\mathbf{v}(j) \cdot \mathbf{u}(k) - \varepsilon} < \varepsilon.$$

The asymptotic factorization of $\mathbf{B}(\mathbf{i}_N)/||\mathbf{B}(\mathbf{i}_N)||$ and the convergence of $\mathbf{u}(\mathbf{i}_N)$ to $\mathbf{u}_\infty(\mathbf{i})$, with $O(\delta^N)$ corrections, follow immediately from the estimates (4.11) and (4.12). □

The reader will of course be curious whether the abstract factorization theorem has any relevance to our one nontrivial example of a recursive tiling tree, namely that of the $\pi/7$ model. For simplicity, we restrict ourselves to the cases where $i_k \in \{1,2\}$, where the matrices $\mathbf{B}(i)$ are identified with $\mathbf{A}(i)$, $i = 0,\ldots,5$. The numerical calculation of the quantities η, ε, c, and δ is easy, but the results are discouraging: δ turns out to be greater than 1, and so the matrices are not well conditioned.

Fortunately, all hope is not lost. The remedy is *telescoping* of the recursion tree. We break up the infinite string into K-element sub-strings, defining new matrices $\mathbf{B}(i)$ as K-fold linked products of $\mathbf{A}(i)$ matrices. The first KN levels of the original tree are now replaced by N levels of the telescoped tree. The reason this might help is our conjecture, backed up by numerical experiments, that long products of $\mathbf{A}(i)$ really do approach factorizability, so that taking products of K matrices should reduce significantly the quantity $r(i)$, and perhaps δ as well. For our $\pi/7$ example, attempts with $K \leq 6$ continue to produce $\delta > 1$, but $K = 7$, with a set of 4374 incidence matrices, is a winner! In Lowenstein (2007), the relevant quantities were calculated exactly, yielding, with 4-digit precision,

$$\eta = 7.297 \times 10^{-6}, \quad \varepsilon = 1.137 \times 10^{-4}, \quad c = 37.60, \quad \delta = 0.6183.$$

4.8 Telescoping

Once asymptotic factorizability has been established, the advantages of further telescoping are considerable and immediate, as follows from the following theorem.

Theorem 4.3. *Given well conditioned matrices* $\mathbf{B}(i)$, $i = 0,\ldots,M-1$, *define the telescoped incidence matrix* $\mathbf{B}_L(k)$, $k = 0,\ldots,M_L - 1$ *to be the kth member of the set*

$$\{\mathbf{B}(\mathbf{i}_L) : i_a = 0,\ldots,M-1,\ n(i_a) = h(i_{a-1})\},$$

listed in numerical order. For the new family of matrices, define $\mathbf{u}_L, \mathbf{v}_L, \mathbf{w}_L, r_L, \boldsymbol{\Delta}\mathbf{u}_L$, $\boldsymbol{\Delta}\mathbf{v}_L$ *as in Definition* 4.1. *Then*

$$r_L(\mathbf{k}_N) \leq \eta_L \delta_L^{N-1},$$

$$|\Delta u_L(\mathbf{k}_N)_j| \leq C\eta_L \delta_L^{N-2},$$

$$|\Delta v_L(\mathbf{k}_N)^j| \leq \varepsilon_L,$$

where

$$\delta_L = \delta^L < 1, \quad \eta_L = \eta\delta^{L-1}, \quad \varepsilon_L = C\eta\delta^{L-1},$$

and

$$C^{-1} = c^{-1} - \frac{c\eta}{1-\delta}(1+\varepsilon).$$

Moreover

$$\frac{\mathbf{B}_L(\mathbf{k}_N)}{||\mathbf{B}_L(\mathbf{k}_N)||} = \mathbf{u}_L(\mathbf{k}_N) \otimes \mathbf{v}_L(\mathbf{k}_N) + O(\delta_L^N),$$

$$|\mathbf{u}_L(\mathbf{k}_N) - \mathbf{u}_\infty(\mathbf{k})| = O(\delta_L^N),$$

and

$$\min_{n(j)=h(i)} \{\mathbf{v}_L(i) \cdot \mathbf{u}_L(j)\} - \varepsilon_L \geq C^{-1}.$$

Proof. The estimate for $r(\mathbf{k}_N)$ follows immediately from (4.11). For Δu_L and Δv_L, we use Theorem 4.2 to obtain:

$$|u(i_{(N-1)L+1}\cdots i_{NL})_k - u(i_{(N-1)L+1})_k| \leq c\eta(\delta^{L-2}+\delta^{L-3}+\cdots+1) \leq \frac{c\eta}{1-\delta} \stackrel{\text{def}}{=} \varepsilon',$$

$$|v(i_1\cdots i_{(N-1)L})^j - v(i_{(N-1)L})^j| \leq \varepsilon.$$

Thus, with the help of (4.15) and (4.16),

$$\begin{aligned}
|\Delta u_L(\mathbf{k}_N)_j| &= |u(\mathbf{i}_{NL})_j - u(\mathbf{i}_{(N-1)L})_j| \leq \frac{r(\mathbf{i}_{(N-1)L})}{\mathbf{v}(\mathbf{i}_{(N-1)L}) \cdot \mathbf{u}(i_{(N-1)L+1}\cdots i_{NL})} \\
&\leq \frac{\eta\delta^{(N-1)L-1}}{\mathbf{v}(i_{(N-1)L}) \cdot \mathbf{u}(i_{(N-1)L+1}) - (\varepsilon+\varepsilon'+\varepsilon\varepsilon')} \leq C\eta_L\delta_L^{N-2},
\end{aligned}$$

$$\begin{aligned}
|\Delta v_L(\mathbf{k}_N)^j| &= |v(\mathbf{i}_{NL})^j - v(i_{(N-1)L}\cdots i_{NL})^j| \leq \frac{r(i_{(N-1)L+1}\cdots i_{NL})}{\mathbf{v}(\mathbf{i}_{(N-1)L}) \cdot \mathbf{u}(i_{(N-1)L+1}\cdots i_{NL})} \\
&\leq \frac{\eta\delta^{L-1}}{\mathbf{v}(i_{(N-1)L}) \cdot \mathbf{u}(i_{(N-1)L+1}) - (\varepsilon+\varepsilon'+\varepsilon\varepsilon')} \leq \varepsilon_L.
\end{aligned}$$

□

4.9 Unique ergodicity for each $\Sigma(\mathbf{i})$

In the models with quadratic irrational parameter, the asymptotic factorization of the products of incidence matrices was closely tied to the existence of a unique invariant measure supported on the closure of the residual set. Recall that the vector $\mathbf{u}$ appearing in the asymptotic tensor product was proportional to an eigenvector of the incidence matrix, which, up to normalization, provided the Hausdorff measures of the various tiles. That a similar situation exists for each of the infinitely many sectors $\Sigma(\mathbf{i})$ is the content of the following theorem of Lowenstein (2007).

Theorem 4.4. *Consider a recursively tiled dressed domain $\Delta^{[n]}$. Let $\mathbf{i}$ be an infinite linked symbol sequence with $n(i_1) = n$, and $\Sigma(\mathbf{i})$ the corresponding invariant component of the residual set. Suppose that the family of $J(n(i)) \times J(h(i))$ incidence matrices $\mathbf{A}(i)$ is well conditioned. Then $\rho^{[n]}$ restricted to $\Sigma(\mathbf{i})$ is uniquely ergodic. The measure on each tile $d_k(\mathbf{i}_L)$ is given, up to renormalization, by the quantity $u_\infty(i_{L+1},\ldots)_k$ appearing in the asymptotic factorization.*

Proof. We begin by defining a positive invariant function $\mu_\mathbf{i}$ on residual tiles of $\Sigma(\mathbf{i})$ by

$$\mu_\mathbf{i}(d_k^{\mathbf{t}_L}(\mathbf{i}_L)) = \mu_\mathbf{i}(d_k(\mathbf{i}_L)) = \frac{u_\infty(i_{L+1}\cdots)_k}{||\mathbf{A}(\mathbf{i}_L)||\mathbf{v}(\mathbf{i}_L)\cdot\mathbf{u}_\infty(i_{L+1}\cdots)}, \tag{4.18}$$

where the denominator has been chosen to enforce the normalization condition

$$\sum_{j,k} A(\mathbf{i}_L)_k^j \mu_\mathbf{i}(d_j(\mathbf{i}_L)) = 1.$$

Further, on any set X which does not intersect the residual tiles, we assign $\mu_\mathbf{i}(X) = 0$. By Proposition 1.7 of Falconer (1990), if we can establish the additivity condition

$$\mu_\mathbf{i}(d_k(\mathbf{i}_L)) = \sum_j A(i_{L+1})_k^j \mu_\mathbf{i}(d_j(\mathbf{i}_{L+1})), \tag{4.19}$$

we will be able to conclude that $\mu_\mathbf{i}$ can be extended to a unique measure on the Borel sets of $\mathbb{R}^2$, with support on the closure of $\Sigma(\mathbf{i})$, and so the theorem will be proved.

Our starting point for the additivity proof is the recursive factorization of multi-level incidence matrices:

$$A(i_{L+1},\ldots,i_N)_k^j = \sum_l A(i_{L+1})_k^l A(i_{L+2},\ldots,i_N)_l^j.$$

Summing over j yields

$$||\mathbf{A}(i_{L+1},\ldots,i_N)||\,u(i_{L+1},\ldots,i_N)_k = \sum_l A(i_{L+1})_k^l ||\mathbf{A}(i_{L+2},\ldots,i_N)||\,u(i_{L+2},\ldots,i_N)_l. \tag{4.20}$$

We now use our asymptotic factorization result, Theorem 4.2, to extract the asymptotic large-N behaviour,

$$\begin{aligned} u(i_{L+1},\ldots,i_N)_k &= u_\infty(i_{L+1},\ldots)_k + O(\delta^N) \\ &= ||\mathbf{A}(i_L)||\mathbf{v}(\mathbf{i}_L)\cdot\mathbf{u}_\infty(i_{L+1},\ldots)\,\mu_\mathbf{i}(d_k(\mathbf{i}_L)) + O(\delta^N), \end{aligned}$$

with a similar expression for $u(i_{L+2},\ldots,i_N)_k$. Inserting these into (4.20) and dividing through by a common factor, we get

$$\begin{aligned} &\frac{\mathbf{v}(\mathbf{i}_L)\cdot\mathbf{u}_\infty(i_{L+1},\ldots)}{\mathbf{v}(\mathbf{i}_L)\cdot\mathbf{u}_(i_{L+1},\ldots,i_N)}\,\mu_\mathbf{i}(d_k(\mathbf{i}_L)) \\ &= \frac{\mathbf{v}(\mathbf{i}_{L+1})\cdot\mathbf{u}_\infty(i_{L+2},\ldots)}{\mathbf{v}(\mathbf{i}_{L+1})\cdot\mathbf{u}(i_{L+2},\ldots,i_N)}\sum_l A(i_{L+1})_k^l \mu_\mathbf{i}(d_l(\mathbf{i}_{L+1})) + O(\delta^N). \end{aligned}$$

In the limit $N \to \infty$, we get (4.19). □

4.10 Multifractal spectrum of recurrence time dimensions

For the quadratic models, there is a pronounced simplicity in the asymptotic scaling, both spatial and temporal, as one proceeds to higher levels of the renormalization. While the characteristic lengths (e.g. diameters of tiles and cells) contract by a factor ω^L, the characteristic time intervals (e.g. periods and return times) grow as a^L, where a is the largest eigenvalue of the incidence matrix. As we saw earlier in this chapter, this makes possible the construction of a transfer matrix $\mathscr{T}(\alpha)$ whose diagonalization determines the Hausdorff dimension (the eigenvalue) and Hausdorff measure (from the eigenvector).

In the present section, we go beyond the quadratic models, making full use of the asymptotic factorization property of incidence matrix products, including the accelerated convergence achieved by telescoping, to describe the asymptotic spatiotemporal properties of the residual set. Instead of a single dimension, characteristic of a fractal, we find a nontrivial continuous spectrum of dimensions characteristic of a *multifractal.*

Before proceeding to our multifractal analysis, we need a more precise definition of the "characteristic time" of a set. Following Afraimovich et al. (2000), we choose the *recurrence time*, defined quite generally as follows:

Definition 4.3. Let S be a set subject to iterated application of a map ϕ. We define the recurrence time $\tau(S)$ as

$$\tau(S) = \min\{t : t \geq 1, \phi^t(S) \cap S \neq \emptyset\}.$$

Within the context of recursive tiling, this notion is particularly apt, since if d_j is a residual base tile, mapped by ρ, then the recurrence time of every tile on the return (to $d(\mathbf{i}_N)$) orbit $\{d^t = \rho^t d_j,\ t = 0, 1, \ldots, T_j\}$ is the same. To see this, we note immediately that $\tau(d^t) \leq \tau(d^0)$, since there are points of d^t whose forward orbits pass through $d^0 \cap d^{\tau(d^0)}$ and return to d^t after a total of $\tau(d^0)$ iterations. On the other hand, the fact that $d^{t+\tau(d^t)} \cap d^t$ is nonempty implies that its preimage $d^{\tau(d^t)} \cap d^0$ is also nonempty, and so $\tau(d^t) \geq \tau(d^0)$. This establishes the equality over the whole return orbit.

We now introduce a generalized dimension which takes into account the joint spatial and temporal scaling properties of subsets of $\Sigma^{[n]}$. We follow the same pattern used to introduce Hausdorff dimension, except that now we have an extra weighting factor equal to a power of the recurrence time. For real q and α, positive δ, and $F \subset \Sigma^{[n]}$, we define

$$\mathscr{M}(\alpha, q, \delta, F) = \inf_{\mathscr{C}_\delta(F)} \sum_{U \in \mathscr{C}_\delta(F)} \tau(U)^{-q} |U|^\alpha,$$

where $\mathscr{C}_\delta(F)$ is a finite or countable covering of F by residual tiles U of ($\mathscr{Q}$-metric) diameter $|U| \leq \delta$, and $\tau(U)$ is the recurrence time for U. For δ tending to zero, the function $\mathscr{M}(\alpha,q,\delta,F)$ is nondecreasing. We set

$$\mathscr{M}(\alpha,q,F) = \lim_{\delta \to 0} \mathscr{M}(\alpha,q,\delta,F),$$

where we allow the possibility $\mathscr{M}(\alpha,q,F) = \infty$.

As a function of α, $\mathscr{M}(\alpha,q,F)$ satisfies:

$$\mathscr{M}(\alpha_0,q,F) = 0 \quad \Longrightarrow \quad \forall \alpha > \alpha_0,\ \mathscr{M}(\alpha,q,F) = 0.$$

$$\mathscr{M}(\alpha_0,q,F) = \infty \quad \Longrightarrow \quad \forall \alpha < \alpha_0,\ \mathscr{M}(\alpha,q,F) = \infty.$$

If $\mathscr{M}(\alpha,q,F)$ is not identically 0 or ∞ as a function of α, then there exists a critical value, the *recurrence time dimension* $\alpha(q,F)$, such that

$$\mathscr{M}(\alpha,q,F) = \begin{cases} \infty & \alpha < \alpha(q,F), \\ 0 & \alpha > \alpha(q,F). \end{cases}$$

The set function $\mathscr{M}(\alpha,q,F)$ is a special case of a Carathéodory outer measure (Pesin, 1998) and $\alpha(q,F)$ is the corresponding Carathéodory dimension.. This concept has wide applicability in multifractal theory (Pesin, 1998), especially where it can be shown that the dimension is independent of F, so that one can write it as $\alpha(q)$. As we shall see below, this is indeed the case for the recurrence time dimension in our models with well conditioned recursive tiling.

4.10.1 Auxiliary measures and dimensions

As in our earlier discussion of Hausdorff dimension, we seek to reduce the calculation of the recurrence time dimension to a transfer matrix eigenvalue problem, since this technique, if it is feasible, quite effectively captures the essence of the recursive tiling structure and at the same time produces a set-independent result. Unfortunately, there is no hope of an easy construction of a transfer matrix $\mathscr{T}(q,\alpha)$, thanks to the complicated nature of the temporal recursion. There is, however, an indirect route based on the asymptotic factorization of incidence matrix products, which was set forth by Lowenstein (2007) and which will be reproduced in the remainder of this chapter.

The first step is find within the recursive tiling formalism suitable proxies for the domain diameter $|D|$ and recurrence time $\tau(D)$ which appear in the function $\mathscr{M}(\alpha,q,\delta,F)$. The relevant domains D are, of course, the residual tiles $d^{\mathbf{t}_N}(\mathbf{i}_N)$. Here we assume that we have L-level telescoped incidence matrices $\mathbf{B}(i)$, omitting, for the sake of notational simplicity, explicit reference to the parameter L except where it plays an essential role.

The choice of proxy for the diameter is fairly obvious (and was already used in the Hausdorff case), namely the scale factor $\omega_L(\mathbf{i}_N)$, since there are positive, level-independent constants κ_- and κ_+ such that

$$\kappa_- \omega(\mathbf{i}_N) \leq |d^{\mathbf{t}_N}(\mathbf{i}_N)| \leq \kappa_+ \omega(\mathbf{i}_N),$$

and the transition from one level to the next is simple multiplicative scaling:

$$\omega(\mathbf{i}_{N+1}) = \omega(\mathbf{i}_N)\omega(i_{N+1}),$$

exactly what we need for constructing a level-by-level transfer matrix.

The natural choice of a proxy for the recurrence time is presumably

$$\nu(\mathbf{i}_N) = ||\mathbf{B}(\mathbf{i}_N)|| = \sum_j T_j(\mathbf{i}_N).$$

Thanks to the asymptotic large N scaling of the return times $T_j(\mathbf{i}_N)$, it is straightforward to find uniform upper and lower bounds for the ratios $\tau(d_L^{\mathbf{t}_N}(\mathbf{i}_N))/\nu(\mathbf{i}_N)$. However, the quantity $\nu(\mathbf{i}_N)$ does not have the kind of product expansion which would make it ideal for inclusion in a transfer matrix. Specifically, from (4.14), we have

$$\nu(\mathbf{i}_N) = (\mathbf{v}(i_1) \cdot \mathbf{u}(i_2))(\mathbf{v}(i_1 i_2) \cdot \mathbf{u}(i_3)) \cdots (\mathbf{v}(i_1 i_2 \dots i_{N-1}) \cdot \mathbf{u}(i_N)) \prod_{k=1}^{N} ||\mathbf{B}(i_k)||.$$

This is not exactly the kind of product expansion we need, namely one involving a string of factors of the form $(\mathbf{v}(i_k) \cdot \mathbf{u}(i_{k+1}))$, but it is close, in the sense that

$$\mathbf{v}(i_1 i_2 \dots i_K) = \mathbf{v}(i_K) + \mathbf{\Delta v}(\mathbf{i}_K), \quad |\Delta v(\mathbf{i}_K)^j| \leq \varepsilon,$$

with $\varepsilon = \varepsilon_L$ given by Theorem 4.3. This suggests a way out of our problem, namely defining a measure and calculating the dimension using a transfer matrix constructed as if we had the desired product structure, then taking the limit $L \to \infty$ to get a handle on the true outer measure and dimension. This is roughly the strategy which we adopt below.

In order to bracket the true large-L asymptotic behavior, we introduce, for $\eta \in \{-, 0, +\}$,

$$\begin{aligned} \nu^{\eta}(\mathbf{i}_N) = & (\mathbf{v}(i_1) \cdot \mathbf{u}(i_2) - \eta\varepsilon)(\mathbf{v}(i_2) \cdot \mathbf{u}(i_3) - \eta\varepsilon) \cdots \\ & (\mathbf{v}(i_{N-1}) \cdot \mathbf{u}(i_N) - \eta\varepsilon) \prod_{k=1}^{N} ||\mathbf{B}(i_k)||. \end{aligned} \tag{4.21}$$

Thanks to the fact that the $u(i)_k$ are positive numbers which sum to unity, we have

$$|\mathbf{\Delta v}(\mathbf{i}_K) \cdot \mathbf{u}(i_{K+1})| \leq \varepsilon,$$

and so we get the bracketing inequalities

$$\mathbf{v}(i_K)\cdot\mathbf{u}(i_{K+1})-\varepsilon\le\mathbf{v}(\mathbf{i}_K)\cdot\mathbf{u}(i_{K+1})\le\mathbf{v}(i_K)\cdot\mathbf{u}(i_{K+1})+\varepsilon,$$

$$v^+(\mathbf{i}_N)\le v(\mathbf{i}_N)\le v^-(\mathbf{i}_N), \tag{4.22}$$

as well as the same estimates with v replaced by v^0.

As intended, the spatial and temporal domain functions now have parallel level-by-level factorization properties, which we can write in the form of scaling ratios,

$$\frac{\omega(\mathbf{i}_N)}{\omega(\mathbf{i}_K)}=\frac{\omega(i_K\cdots i_N)}{\omega(i_K)},\qquad \frac{v^\eta(\mathbf{i}_N)}{v^\eta(\mathbf{i}_K)}=\frac{v^\eta(i_K\cdots i_N)}{v^\eta(i_K)} \tag{4.23}$$

Now we are in a position to define measures μ^η which will bracket the recurrence-time outer measure and produce, via the construction and diagonalization of transfer matrices, approximants to the recurrence-time dimension $\alpha(q)$. To this end, we introduce the following positive invariant function on residual tiles:

$$\begin{aligned}\mu^\eta(\alpha,q,L,N,d_j^{\mathbf{t}_K}(\mathbf{i}_K)) &= \mu^\eta(\alpha,q,L,N,d_j(\mathbf{i}_K))\\ &= \sum_{k,i_{K+1},\ldots,i_N} B(i_{K+1},\ldots,i_N)_j^k\, v^\eta(\mathbf{i}_N)^{-q}\,\omega(\mathbf{i}_N)^\alpha, \qquad (4.24)\\ \mu^\eta(\alpha,q,L,N,d_j^{[n]}) &= \sum_{\mathbf{i}_N,k} B(\mathbf{i}_N)_j^k v^\eta(\mathbf{i}_N)^{-q}\omega(\mathbf{i}_N)^\alpha.\end{aligned}$$

We note that N plays the role of δ in the definition of $\mathscr{M}(\alpha,q,\delta,F)$, with $N\to\infty$ corresponding to $\delta\to 0$, and recall that the factor $B(\mathbf{i}_N)_j^k$ counts the number of level-N residual tiles $d_k^{\mathbf{t}_N}(\mathbf{i}_N)$ which cover the level-K base tile $d_j^{\mathbf{i}_K}(\mathbf{i}_K)$. At this stage, the telescoping depth L remains an adjustable parameter.

The extension of μ^η to finite and countable unions of disjoint tiles is immediate, employing the easily established additivity formulas

$$\begin{aligned}\mu^\eta(\alpha,q,L,N,d_j(\mathbf{i}_K)) &= \sum_{i_{K+1},k}\delta_{h(i_K),n(i_{K+1})}B(i_{K+1})_j^k\mu^\eta(\alpha,q,L,N,d_k(\mathbf{i}_{K+1})),\\ \mu^\eta(\alpha,q,L,N,d_j^{[n]}) &= \sum_{i,k}\delta_{n,n(i)}B(i)_j^k\mu^\eta(d_k(i)).\end{aligned} \tag{4.25}$$

In addition, the scaling relation

$$\frac{\mu^\eta(\alpha,q,L,N,d_j(\mathbf{i}_K))}{v^\eta(\mathbf{i}_K)^{-q}\,\omega(\mathbf{i}_K)^\alpha}=\frac{\mu^\eta(\alpha,q,L,N,d_j(i_K))}{v^\eta(i_K)^{-q}\,\omega(i_K)^\alpha}. \tag{4.26}$$

is a straightforward consequence of the factorization properties (4.23).

We now introduce the doubly indexed transfer matrix

$$\mathscr{T}^\eta(\alpha,q,L)_{(i_1,j_1),(i_2,j_2)}=\delta_{h(i_1),n(i_2)}B(i_2)_{j_1}^{j_2}\left(||\mathbf{B}(i_1)||(\mathbf{v}(i_1)\cdot\mathbf{u}(i_2)-\eta\varepsilon)\right)^{-q}\omega(i_2)^\alpha. \tag{4.27}$$

It is not difficult to write $\mu^\eta(\alpha,q,L,N,d_j(\mathbf{i}_K))$ as a matrix element of this matrix between N-independent vectors Φ^η and Ψ_j^η, with components

$$\Phi^{\eta}(q,L)_{(i,j)} = v^{\eta}(i)^{-q},$$

$$\Psi_j^{\eta}(\alpha,q,L,\mathbf{i}_K)_{(i',j')} = \delta_{jj'}\omega(\mathbf{i}_K)\begin{cases} 1 & K=1 \\ (v^{\eta}(\mathbf{i}_K)/v^{\eta}(i_K))^{-q} & K>1. \end{cases}$$

Two theorems from Lowenstein (2007) complete the extraction of the recurrence-time dimension. The first of these obtains the approximants $\alpha(q,L)$ from the transfer-matrix eigenvalue problem:

Theorem 4.5. (*Lowenstein,* 2007, *Theorem* 9) *Suppose that for all real q and α, $\eta \in \{-,0,+\}$, and positive integer L, there exists a positive integer n such that all elements of the matrix $\mathscr{T}^{\eta}(\alpha,q,L)^n$ are strictly positive. Then, for every tile F, there exists a real number $\alpha^{\eta}(q,L)$ such that*

$$\lim_{N\to\infty} \mu^{\eta}(\alpha,q,L,N,F) = \begin{cases} \infty & \alpha < \alpha^{\eta}(q,L), \\ \mu^{\eta}(\alpha,q,L,F) & \alpha = \alpha^{\eta}(q,L), \\ 0 & \alpha > \alpha^{\eta}(q,L), \end{cases}$$

where $\mu^{\eta}(\alpha,q,L,F)$ has the same additivity and scaling properties as $\mu^{\eta}(\alpha,q,L,N,F)$ and can be extended to a measure on the Borel sets of $\mathbb{R}^2$. The three dimensions satisfy the inequalities

$$\alpha^{-\mathrm{sign}(q)}(q,L) \leq \alpha^{0}(q,L) \leq \alpha^{\mathrm{sign}(q)}(q,L).$$

Proof. The Perron-Frobenius theorem (see Katok and Hasselblatt, 1995) tells us that the transfer matrix has an isolated largest eigenvalue $\lambda^{\eta}(\alpha,q,L)$ associated with a normalized eigenvector $\Omega^{\eta}(\alpha,q,L)$ with all components positive. Since every element of the transfer matrix depends on α only through a factor ω^{α} with $0<\omega<1$, we infer that $\lambda^{\eta}(\alpha,q,L)$ decreases monotonically from $+\infty$ to 0 as α ranges over $(-\infty,+\infty)$, passing through the value 1 for unique $\alpha^{\eta}(q,L)$. This gives the stated behavior of μ^{η} for $N\to\infty$, with $\mu^{\eta}(\alpha^{\eta}(q,L),q,L,F)$ a finite, nonzero matrix element of the projector along $\Omega^{\eta}(\alpha^{\eta}(q,L),q,L)$. The additivity and scaling properties are immediate consequences of (4.25) and (4.26), and the extension to a measure follows from Proposition 1.7 of Falconer (1990). The inequalities for the dimensions are a direct consequence of (4.22) and the monotonic decrease of all transfer matrix elements as functions of α. □

The second theorem of Lowenstein (2007) establishes the connection to the recurrence-time dimension in the limit of infinite L.

Theorem 4.6. (*Lowenstein,* 2007, *Theorem* 10) *For a telescoped recursive tiling scheme with L-fold incidence matrices satisfying the conditions of Theorems* 4.2 *and* 4.3, *and a transfer matrix satisfying the hypothesis of Theorem* 4.5, *the dimensions $\alpha^{\eta}(q,L)$, $\eta \in \{-,0,+\}$ have a common $L\to\infty$ limit, which coincides with a tile-independent recurrence-time dimension $\alpha(q)$.*

Proof. To show that the three dimensions $\alpha^{\eta}(q,L)$ differ by amounts which vanish in the $L \to \infty$ limit, we focus on the μ^{η} measures of the lowest level tiles $d^{[n]}$ and assume $q > 0$. The case $q < 0$ is strictly analogous and for $q = 0$, the Hausdorff case, the three dimensions are identical for all L. We write

$$\begin{aligned}\mu^{+}(\alpha,q,N,d^{[n]}) &= \Sigma_{\mathbf{i}_N} \nu(\mathbf{i}_N)\nu^{+}(\mathbf{i}_N)^{-q}\omega(\mathbf{i}_N)^{\alpha} \\ &= \Sigma_{\mathbf{i}_N} \nu(\mathbf{i}_N)\nu^{-}(\mathbf{i}_N)^{-q}\omega(\mathbf{i}_N)^{\alpha^{-}(q,L)} \left(\tfrac{\nu^{+}(\mathbf{i}_N)}{\nu^{-}(\mathbf{i}_N)}\right)^{-q} \omega(\mathbf{i}_N)^{\alpha-\alpha^{-}(q,L)}.\end{aligned}$$

But, from (4.21),

$$1 \le \left(\frac{\nu^{+}(\mathbf{i}_N)}{\nu^{-}(\mathbf{i}_N)}\right)^{-q} = \prod_{a=1}^{N-1}\left(\frac{\nu(i_a)\cdot u(i_{a+1}) - \varepsilon}{\nu(i_a)\cdot u(i_{a+1}) + \varepsilon}\right)^{-q} \le (1+2C\varepsilon)^{qN}$$

and

$$\omega_{\min}^{LN} \le \omega(\mathbf{i}_N) \le \omega_{\max}^{LN},$$

where the L-independent constant C is defined in Theorem 4.3, and $\omega_{\min} = \min\{\omega(i)\}$ and $\omega_{\max} = \max\{\omega(i)\}$ are independent of L. Thus

$$\mu^{+}(\alpha,q,N,d^{[n]}) \le \mu^{-}(\alpha^{-}(q,L),q,N,d^{[n]})H(q,L)\left((1+2C\varepsilon)^{q}\,\omega_{\max}^{L(\alpha-\alpha^{-}(q,L))}\right)^{N},$$

so that if

$$\alpha - \alpha^{-}(q,L) > \frac{q\log(1+2C\varepsilon)}{L\log(\omega_{\max}^{-1})} \stackrel{\text{def}}{=} \xi(q,L),$$

then

$$\mu^{+}(\alpha,q,L,d^{[n]}) = \lim_{N\to\infty}\mu^{+}(\alpha,q,L,N,d^{[n]}) = 0.$$

It follows that

$$0 \le \alpha^{+}(q,L) - \alpha^{-}(q,L) \le \xi(q,L),$$

since otherwise we could find an α between $\alpha^{-}(q,L)+\xi(q,L)$ and $\alpha^{+}(q,L)$ such that $\mu^{+}(\alpha,q,L,d^{[n]}) = 0$, in contradiction to the definition of $\alpha^{+}(q,L)$. But $\lim_{L\to\infty}\xi(L) = 0$, and hence the three dimensions have a common limit.

In the final segment of the proof, we show that the recurrence-time dimension $\alpha(q)$ is sandwiched between $\alpha^{-}(q,L)$ and $\alpha^{+}(q,L)$ for all L, and hence coincides with their common limit when L tends to infinity. Once again making use of the level-independent upper and lower bounds on the ratios $\tau(d)/\nu(d)$ and $|d|/\omega(d)$ for arbitrary residual tiles d, as well as the bracketing of $\nu(d)$ between $\nu^{+}(d)$ and $\nu^{-}(d)$, we know there exist positive coefficients $C_{\pm}(\alpha,q,L)$ such that

$$\mathscr{M}(\alpha,q,\delta.F) \ge C_{-}(\alpha,q,L)\inf_{\mathscr{C}_{\delta}(F)}\sum_{d\in\mathscr{C}_{\delta}(F)}\nu^{-}(d)^{-q}\omega(d)^{\alpha},$$

$$\mathscr{M}(\alpha,q,\delta.F) \le C_{+}(\alpha,q,L)\inf_{\mathscr{C}_{\delta}(F)}\sum_{d\in\mathscr{C}_{\delta}(F)}\nu^{+}(d)^{-q}\omega(d)^{\alpha}.$$

But the scaling relation (4.26) implies the bounds

$$\nu^-(d)^{-q}\omega(d)^\alpha \geq \mu^-(\alpha,q,L,d)\min_{i,j}\frac{\nu^-(i)^{-q}\omega(i)^\alpha}{\mu^-(\alpha,q,L,d_j(i))},$$

$$\nu^+(d)^{-q}\omega(d)^\alpha \leq \mu^+(\alpha,q,L,d)\max_{i,j}\frac{\nu^-(i)^{-q}\omega(i)^\alpha}{\mu^-(\alpha,q,L,d_j(i))},$$

and so there exist positive coefficients $K_\pm(\alpha,q,L)$ such that

$$\mathscr{M}(\alpha,q,\delta.F) \geq K_-(\alpha,q,L)\inf_{\mathscr{C}_\delta(F)}\sum_{d\in\mathscr{C}_\delta(F)}\mu^-(\alpha,q,L,d)$$

$$\mathscr{M}(\alpha,q,\delta.F) \leq K_+(\alpha,q,L)\inf_{\mathscr{C}_\delta(F)}\sum_{d\in\mathscr{C}_\delta(F)}\mu^+(\alpha,q,L,d).$$

The inequalities imply that $\mathscr{M}(\alpha,q,\delta.F)$ is infinite for $\alpha < \alpha^-(q,L)$ and vanishes for $\alpha > \alpha^+(q,L)$, and hence the critical value $\alpha(q,F)$ is caught between those two F-independent values for all L. In the limit, they all coincide and are independent of F. □

4.10.2 Simpler calculation of the recurrence time dimensions

Theorems 4.5 and 4.6 suggest a straightforward algorithm for a numerical calculation of the recurrence time dimension $\alpha(q)$: Calculate the largest eigenvalue $\lambda(\alpha,q,L)$ of the transfer matrix $\mathscr{T}^0(\alpha,q)$ and use standard root-finding to determine the value of α where $\lambda(\alpha,q,L) = 1$. For fixed q, repeat the experiment with increasingly large L until the desired precision has been attained. Lowenstein (2007) found that a simpler eigenvalue problem, using the single-index matrix

$$\hat{\mathscr{T}}(\alpha,q,L)_{i_1,i_2} = \omega(i_1)^\alpha(||B(i_1)||v(i_1)\cdot u(i_2))^{1-q},$$

produces the same dimension as the one based on $\mathscr{T}^0(\alpha,q,L)$ defined in (4.27). To prove this, it is not necessary to use the simpler transfer matrix to construct a full-fledged measure. In fact, that may not be possible. Rather, we need only concentrate on a specific level-0 base tile, $d^{[n]}$, defining

$$\hat{\mu}(\alpha,q,L,N,d^{[n]}) = \sum_{\mathbf{i}_N}\nu^0(\mathbf{i}_N)^{1-q}\omega(\mathbf{i}_N)^\alpha = \sum_{i_1,i_N}\hat{\mathscr{T}}(\alpha,q,L)^{N-1}_{i_1,i_N}\nu^0(i_N)^{1-q}\omega(i_N)^\alpha.$$

Now we can use the same strategy which was used in the first part of the proof of Theorem 4.6, writing

$$\hat{\mu}(\alpha,q,N,d^{[n]}) = \sum_{\mathbf{i}_N}\nu(\mathbf{i}_N)\nu^0(\mathbf{i}_N)^{-q}\omega(\mathbf{i}_N)^{\alpha^0(q,L)}\frac{\nu^0(\mathbf{i}_N)}{\nu(\mathbf{i}_N)}\omega(\mathbf{i}_N)^{\alpha-\alpha^0(q,L)}.$$

But

$$\frac{\nu(\mathbf{i}_N)}{\nu^0(\mathbf{i}_N)} = \prod_{a=1}^{N-1} \left(\frac{\mathbf{v}(\mathbf{i}_a) \cdot \mathbf{u}(i_{a+1})}{\mathbf{v}(i_a) \cdot \mathbf{u}(i_{a+1})} \right)$$

with each factor bounded above and below as

$$(1+\varepsilon C)^{-1} \le \left(\frac{\mathbf{v}(\mathbf{i}_a) \cdot \mathbf{u}(i_{a+1})}{\mathbf{v}(i_a) \cdot \mathbf{u}(i_{a+1})} \right) \le (1+\varepsilon C)$$

Thus

$$(1+\varepsilon C)^{-N+1} \omega_{\min}^{NL(\alpha-\alpha^0(q,L))} \le \frac{\hat{\mu}(\alpha,q,N,d^{[n]})}{\mu^0(\alpha,q,N,d^{[n]})} \le (1+\varepsilon C)^{N-1} \omega_{\max}^{NL(\alpha-\alpha^0(q,L))}.$$

Defining

$$\hat{\mu}(\alpha,q,L,d^{[n]}) = \lim_{N\to\infty} \hat{\mu}(\alpha,q,L,N,d^{[n]}),$$

we have

$$\alpha - \alpha^0(q,L) > \frac{\log(1+\varepsilon C)}{L\log(\omega_{\max}^{-1})} \stackrel{\text{def}}{=} \xi_2(q,L) \Longrightarrow \hat{\mu}(\alpha,q,L,d^{[n]}) = 0$$

and

$$\alpha - \alpha^0(q,L) < -\frac{\log(1+\varepsilon C)}{L\log(\omega_{\min}^{-1})} \stackrel{\text{def}}{=} -\xi_1(q,L) \Longrightarrow \hat{\mu}(\alpha,q,L,d^{[n]}) = \infty,$$

and so the critical α value $\hat{\alpha}(q,L)$ lies in the interval

$$\alpha^0(q,L) - \xi_1(q,L) \le \hat{\alpha}(q,L) \le \alpha^0(q,L) + \xi_2(q,L)$$

for all L, and hence converges to $\alpha(q)$ in the limit.

The reformulation of the eigenvalue problem allows us to write down immediately the recurrence-time dimension for renormalizable models with unique spatial and temporal scale factors, $\omega < 1$ and $\tau > 1$. This includes not only our family of quadratic models, but also those individual ergodic components of the cubic $\pi/7$ model which are characterized by a repeating sequence $\mathbf{i} = (i_1, i_2, \ldots, i_M)^\infty$. In the latter case, τ is the largest eigenvalue of the M-fold telescoped incidence matrix, and

$$\omega = \prod_{m=1}^{M} \omega(i_m).$$

It is clear that for finite N, $\hat{\mu}$ is proportional to

$$\sum_{\mathbf{i}_N} \tau^{1-q} \omega^{\alpha},$$

which tends to 0 (resp.∞) for $\alpha > (q-1)\log\tau/\log\omega$ (resp. $\alpha < (q-1)\log\tau/\log\omega$. Thus the recurrence-time spectrum of dimensions is just the linear function

$$\alpha(q) = (1-q)D, \qquad D = -\frac{\log \tau}{\log \omega}. \tag{4.28}$$

We note that $\alpha(0) = D$ is just the familiar Hausdorff dimension.

4.10.3 Recurrence time spectrum for the $\pi/7$ model

The residual set of the $\pi/7$ model is characterized by infinitely many spatial and temporal scale factors, and so we do not expect a closed-form expression for the dimension spectrum $\alpha(q)$. However, the transfer matrix method makes it straightforward to obtain quite accurate numerical values with a modest input of computational effort. As we found in the calculation of Hausdorff dimension earlier in this chapter, the structure of the transfer matrix is such that the eigenvalue problem factorizes and we need only consider the two-member catalogue $\{\Delta^{[1]}, \Delta^{[2]}\}$.

Using the incidence matrices for the catalogue, namely $\mathbf{A}(i)$, $i = 0,\ldots,5$ listed in Tables A.5 and A.6, we calculated the coefficients of $\hat{\mathscr{T}}(\alpha,q,L)$ for telescoping depths $L = 5,6,7$. For each of 30 equally spaced values of q in the range $0 \le q \le 1.45$, we calculated (Lowenstein, 2007) the largest eigenvalue of the transfer matrix by the method of iteration, homing in on $\hat{\alpha}(q,L)$ by means of the secant method of root-finding. The results for $L = 7$ are given in Table 4.1, where we have kept only those digits which are judged conservatively (from a comparison with the $L = 6$ numbers) to have converged to their infinite-L values. The range of q values was extended to $-8 \le q \le 8$, using the $L = 6$ matrix, in order to obtain the graph of Fig. 4.1.

Table 4.1 Selected $\hat{\alpha}^0(q,L)$ values for $L = 7$.

q	$\hat{\alpha}_7^0(q)$	q	$\hat{\alpha}_7^0(q)$	q	$\hat{\alpha}_7^0(q)$
0	1.652233652	0.5	1.047005707	1	0.4602540423
0.05	1.591027785	0.55	0.9874103750	1.05	0.4028042538
0.1	1.529960826	0.6	0.9280048043	1.1	0.3455986983
0.15	1.469037424	0.65	0.8687944874	1.15	0.2886424095
0.2	1.408262337	0.7	0.8097849509	1.2	0.2319402547
0.25	1.347640428	0.75	0.7509817413	1.25	0.1754969094
0.3	1.287176666	0.8	0.6923904094	1.3	0.1193168306
0.35	1.226876115	0.85	0.6340164934	1.35	0.06340423039
0.4	1.166743930	0.9	0.5758655008	1.4	0.007763050387
0.45	1.106785355	0.95	0.5179428886	1.45	−0.04760306470

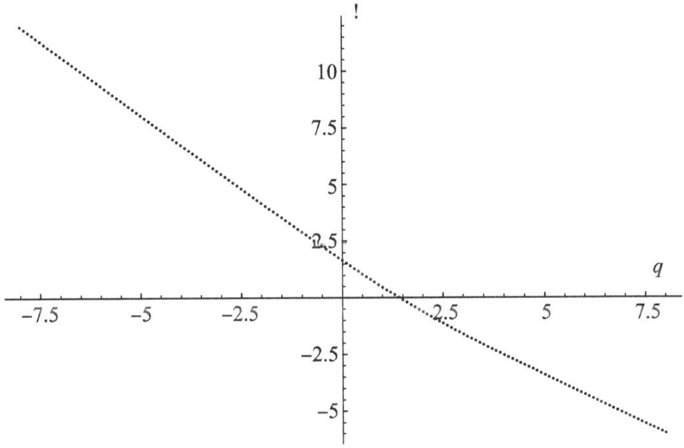

Fig. 4.1 The $\alpha(q)$ curve for Example II.

It is instructive to compare the values of $\hat{\alpha}(q,7)$, which we believe to be excellent approximations to $\alpha(q)$, with values obtained by independent methods at three points on the curve: $q=0$, $q=1$, and the point $q=q_0$ such that $\alpha(q_0)=0$. From the definition of $\mathscr{M}(\alpha,q,d^{[n]})$, $\alpha(0)$ is the Hausdorff dimension of $d^{[n]}$, while $\alpha(1)$ is that of the fractal set obtained by taking the $N \to \infty$ limit of the union over all base tiles $d(\mathbf{i}_N)$ with $n(i_1)=n$. Earlier in this chapter, the transfer matrix method was used to obtain $\alpha(0)$ as the solution of a transcendental equation of the form

$$F(\eta_1^{\alpha(0)}, \eta_2^{\alpha(0)}) = 0,$$

where η_1 and η_2 are the fundamental units of the ring $\mathbb{Z}[\lambda]$ and $F(x,y)$ is a polynomial of degree 14. This yielded the value

$$\alpha(0) = 1.65223365188166270 81\ldots.$$

The analogous calculation for the base-tile fractal gives $\alpha(1)$ as the solution of

$$G(\eta_1^{\alpha(1)}, \eta_2^{\alpha(1)}) = 0,$$

where

$$G(x,y) = x^2y^2 + xy^3 + y^2 - 1.$$

A numerical value with 20-digit precision is easily obtained by Newton's method:

$$\alpha(1) = 0.46025404225607400229\ldots$$

It is encouraging that the $L=7$ numerical transfer matrix calculation reproduces these dimensions with 10-digit precision.

The third point of comparison is less precise, but important for the interpretation of the model as pseudochaotic, i. e. having a power-law increase of complexity over time. From Table 4.1, we see that q_0 is slightly larger than 1.40. Using the same numerical construction of $\hat{\alpha}(q,7)$ as before, supplemented by the secant method to home in on the zero of that function, we obtained

$$q_0 = 1.40699563 \tag{4.29}$$

with an error bounded by 10^{-8}. To interpret this exponent (nonrigorously), we note that the calculation with α set equal to zero parallels closely that of Hausdorff or box-counting dimension, provided that we think of the inverse of the recurrence time as the "size" of a covering set. Then t^{q_0} is seen to be proportional, for asymptotically large t, to the number of tiles in all the return orbits with recurrence time less than or equal to t, a quantity which is one of various possibilities for measuring the growth of complexity in this class of models. Another is the number of polygonal cells of periodic orbits with return time less than t. A numerical experiment by Lowenstein et al. (2004) produced a value for $q_0 - 1$ which agrees with (4.29) to 4 digits of precision.

We note in passing that application of the above calculation of the complexity-growth exponent q_0 to the quadratic cases, using (4.28), gives $q_0 = 1$. This is consistent with the observation that in the quadratic models the number of periodic cells increases, on average, proportional to the return time, as one ascends the renormalization ladder of each scaling sequence.

The plot in Fig. 4.1 suggests that the function $\alpha(q)$ resembles a hyperbola, with linear asymptotes for $q \to \pm\infty$. This type of behavior should not come as a surprise. After all, for $q \to \infty$, the calculation emphasizes the residual tiles with the largest diameters and smallest return times, which are found to scale with a temporal factor $\tau_+ = 4$ and spatial factor $\omega = 2 - \lambda$, where λ is the parameter $2\cos\pi/7$. On the other hand, for $q \to -\infty$, the tiles which are dominant are those with the smallest diameters and longest return times, with scaling factors $\tau_- = \frac{1}{2}(51 + \sqrt{2641})$ and $\omega_- = -5 + \lambda + \lambda^2$. Thus

$$\alpha(q) \sim (1-q)\beta_\pm, \quad q \to \pm\infty,$$

where

$$\beta_+ = -\frac{\log 4}{\log(2-\lambda)} = 0.856\ldots, \qquad \beta_- = -\frac{\log(51+\sqrt{2641})}{2\log(2-\lambda)} = 1.304\ldots.$$

It is sometimes useful to reformulate the description of a multifractal in terms of a Legendre transform of $\alpha(q)$ (see Falconer, 1990, Chap. 17). One advantage of this is that the infinitely extended asymptotes of the $\alpha(q)$ get mapped into the endpoints of a finite curve, whose peak value is just the Hausdorff dimension. Specifically, we define $f(\beta)$ parametrically by the equations

$$\beta(q) = -\frac{d\alpha(q)}{dq},$$
$$f(q) = \alpha(q) + q\beta(q).$$

and display its graph in Fig. 4.2. The interpretation of this function is roughly the following: if one covers the residual set by tiles of diameter approximately equal to some small δ, the number of such tiles with box-counting dimension β will be proportional to $\delta^{-f(\beta)}$.

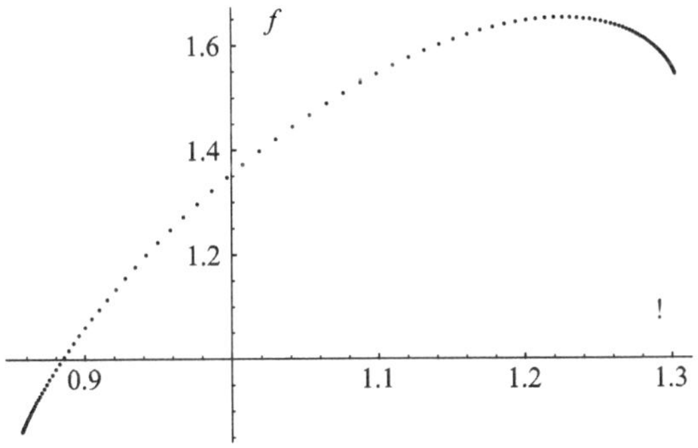

Fig. 4.2 The $f(\beta)$ curve for Example II.

References

Afraimovich V., Schmeling J., Ugalde E., and Urias J. (2000) Spectra of dimensions for Poincaré recurrences, *Discrete and Continuous Dynamical Systems* **6**, 901–914.

Falconer K. (1990) *Fractal Geometry*, Wiley, Chichester.

Katok A. and Hasselblatt B. (1995) *Introduction to the Modern Theory of Dynamical Systems*, Cambridge University Press, Cambridge.

Lowenstein J.H. (2007) Aperiodic orbits of piecewise rational rotations of convex polygons with recursive tiling, *Dynamical Systems* **22**, 25–63.

Lowenstein J.H., Kouptsov K. L., and Vivaldi F. (2004) Recursive tiling and geometry of piecewise rotations by $\pi/7$, *Nonlinearity* **17**, 1–25.

Pesin Ya. B. (1998) *Dimension Theory in Dynamical Systems: Contemporary Views and Applications*, University Chicago Press, Chicago.

Chapter 5
Global Dynamics

5.1 Global expansivity

In the preceding chapters, we have achieved a detailed understanding, based on the organizing principles of renormalization and symbolic coding, of the local dynamics associated with renormalizable kicked-oscillator maps. We now wish to show how the whole recursive return map formalism can be extended ("lifted") to the aperiodic orbits on the infinite plane of the global map W. The local scaling, characterized by the self-similar contraction of domains in the fundamental cell Ω, will be seen to have, if certain conditions are satisfied, a global counterpart in the form of asymptotic self-similar expansion. Just as the local contraction (within a single ergodic component) is associated with a scale factor $\omega_K < 1$, the global expansivity will have its own scale factor $\omega_W > 1$. Given that the temporal behavior scales with a factor ω_T, it is not surprising that orbits will be found to tend to infinity for asymptotically long times with a power-law exponent $\log \omega_W / \log \omega_T$.

5.1.1 *Lifting the return map* $\rho_K(0)$

We recall from Chap. 1 that the sawtooth kicked-oscillator map W can be factored into local and global piecewise linear maps. It is convenient to break up the plane into translates of the square Ω, each such cell labeled by a Gaussian integer $n + mi$, $i = \sqrt{-1}$, $m, n \in \mathbb{Z}$. Each point x of the plane is then represented by a pair $[\mathbf{u}, \zeta]$, with the local point u a point of Ω and $\zeta \in \mathbb{Z}[i]$ labeling the cell. The map W is then given by

$$W[\mathbf{u}, \zeta] = [K(\mathbf{u}), i\zeta + \Delta(\mathbf{u})],$$

the equivalent, in the new notation, of (1.7). The orbit of $[\mathbf{u}, \zeta]$ then consists of the points

$$W^n[\mathbf{u}, \zeta] = [K^n(\mathbf{u}), i^n\zeta + i^{n-1}\Delta(\mathbf{u}) + i^{n-2}\Delta(K(\mathbf{u})) + \cdots + \Delta(K^{n-1}(\mathbf{u}))].$$

Now suppose that $\mathbf{u}$ is a point in one of the domains (level=0 base tiles) $D_j(i_0)$ of a recursive tiling hierarchy, i.e. performing $T_j(i_0)$ iterations of K on $\mathbf{u}$ will map the latter back into $D(i_0)$ for the first time, i.e.

$$K^{T_j(i_0)}(\mathbf{u}) = \rho_K(i_0)(\mathbf{u}).$$

Here and in what follows we will use ρ_K, in place of ρ, to denote the local return map of Sect. 4. Then the same number of W iterations on $[\mathbf{u}, \zeta]$ produces

$$W^{T_j(i_0)}[\mathbf{u}, \zeta] = [\rho_K(i_0)(\mathbf{u}), i^{T_j(i_0)}\zeta + d_j(i_0)] \stackrel{\text{def}}{=} \rho_W(i_0)[\mathbf{u}, \zeta],$$

where $d_j(i_0)$ is the lattice translation associated with the map $\rho_W(i_0)$, namely,

$$d_j(i_0) = i^{T_j(i_0)-1}\Delta(\mathbf{u}) + i^{T_j(i_0)-2}\Delta(K(\mathbf{u})) + \cdots + \Delta(K^{T_j(i_0)-1}(\mathbf{u})).$$

One must be careful here to interpret the notation properly: $\rho_W(i_0)$ is not in general a true first return map, but rather a first return map *modulo a lattice translation*: the domain $\rho_W(i_0)[D_j(i_0), \zeta]$ is identical to $[\rho_K D_j(i_0), \zeta]$, translated to the cell $i^{T_j(i_0)}\zeta + d_j(i_0)$. This is illustrated in Fig. 5.1, which shows the respective local and global orbits of the tile $D_0(0)$ of the fundamental cell in the quadratic Model I with

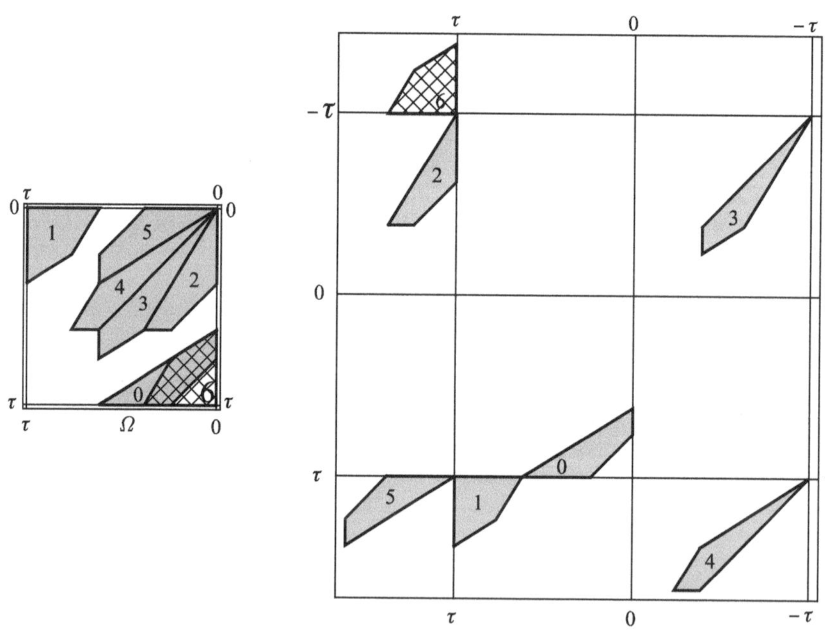

Fig. 5.1 Local and global orbits of the level-0 atom $D_0(0)$. After 6 iterations of the local map K, the image domain has returned, 'rotated', to $D(0)$. After 6 iterations of W, on the other hand, the image domain lands, with the same 'rotation', in $D(0) + (\tau, -2\tau)$.

parameter $\frac{1}{2}(1+\sqrt{5})$. In the local orbit, 6 copies of the tile, rotated as well as translated, are located throughout Ω, sharing with the orbits of $D_1(0)$ and $D_2(0)$ that part of Ω not occupied by periodic pentagons (see Sect.2.3). The global orbit, on the other hand, consists of translates of the 6 tiles located in a number of nearby lattice cells.

We see here the possibility of long-distance transport on the plane under iteration of ρ_W. To investigate this possibility, we must first study the lifted versions of the higher-level return maps $\rho_K(i_0, i_1, \dots, i_L)$.

5.1.2 Lifting the higher-level return maps

To construct the higher-level maps $\rho_W(\mathbf{i}_L)$ which elevate $\rho_K(\mathbf{i}_L)$ to the plane, we proceed recursively. Suppose we have succeeded in defining

$$\rho_W(\mathbf{i}_L)[\mathbf{u},\zeta] = [\rho_K(\mathbf{i}_L)\mathbf{u}, i^{T_j(\mathbf{i}_L)}\zeta + d_j(\mathbf{i}_L)], \quad \mathbf{u} \in D_j(\mathbf{i}_L). \tag{5.1}$$

We wish to construct, for $u \in D_j(\mathbf{i}_{L+1})$,

$$\rho_W(\mathbf{i}_{L+1})\,[\mathbf{u},\zeta] = \rho_W(\mathbf{i}_L)^{\nu(i_{L+1},j)}[\mathbf{u},\zeta],$$

iterating $\rho_W(\mathbf{i}_L)$ along the $\nu(i_{L+1},j)$-step return path $p(\mathbf{i}_{L+1},j)$.

For an arbitrary t-step path $(j_0, j_1, \dots, j_{t-1})$, we have

$$\begin{aligned}
\rho_W(\mathbf{i}_L)[\mathbf{u},\zeta] &= [\rho_K(\mathbf{i}_L)\mathbf{u}, i^{T_{j_0}(\mathbf{i}_L)}\zeta + d_{j_0}(\mathbf{i}_L)],\\
\rho_W(\mathbf{i}_L)^2[\mathbf{u},\zeta] &= [\rho_K(\mathbf{i}_L)^2\mathbf{u}, i^{T_{j_0}(\mathbf{i}_L)+T_{j_1}(\mathbf{i}_L)}\zeta + i^{T_{j_1}(\mathbf{i}_L)}d_{j_0}(\mathbf{i}_L) + d_{j_1}(\mathbf{i}_L)],\\
&\;\;\vdots\\
\rho_W(\mathbf{i}_L)^t[\mathbf{u},\zeta] &= [\rho_K(\mathbf{i}_L)^t\mathbf{u}, i^{\sum_{s=0}^{t-1}T_{j_0}(\mathbf{i}_L)}\zeta + i^{\sum_{s=1}^{t-1}T_{j_0}(\mathbf{i}_L)}d_{j_0}(\mathbf{i}_L) + \cdots + d_{j_{t-1}}(\mathbf{i}_L)].
\end{aligned}$$

Introducing the more compact notation

$$\mathbf{j} = (j_0, \dots, j_{t-1}), \quad T_{\mathbf{j}}(\mathbf{i}_L) = \sum_{s=0}^{t-1} T_{j_s}(\mathbf{i}_L),$$

$$T_{(\mathbf{j}|r)}(\mathbf{i}_L) = \sum_{s=r}^{t-1} T_{j_s}(\mathbf{i}_L), \quad 0 \le r \le t-1, \quad T_{(\mathbf{j}|t)}(\mathbf{i}_L) = 0, \tag{5.2}$$

$$c_{k,\mathbf{j}}(\mathbf{i}_L) = \sum_{s=0}^{t-1} \delta_{k,j_s}\, i^{T_{(\mathbf{j}|s+1)}(\mathbf{i}_L)},$$

we get

$$\rho_W(\mathbf{i}_L)^t[\mathbf{u},\zeta] = [\rho_K(\mathbf{i}_L)^t\,\mathbf{u},\; i^{T_{\mathbf{j}}(\mathbf{i}_L)}\zeta + \sum_k c_{k,\mathbf{j}}(\mathbf{i}_L)\,d_k(\mathbf{i}_L)]. \tag{5.3}$$

For the full return path of $\mathbf{u} \in D_j(\mathbf{i}_{L+1})$, we have $t = \nu(i_L, j)$ and thus, from (5.3),

$$\rho_W(\mathbf{i}_{L+1})[\mathbf{u}, \zeta] = [\rho_K(\mathbf{i}_{L+1})\,\mathbf{u},\; i^{T_{p(i_{L+1},j)}(\mathbf{i}_L)}\zeta + d_j(\mathbf{i}_{L+1})],$$

with

$$d_j(\mathbf{i}_{L+1}) = \sum_K M_{jk}(\mathbf{i}_{L+1})\, d_k(\mathbf{i}_L),$$

$$M_{jk}(\mathbf{i}_{L+1}) = c_{k,p(i_{L+1},j)}(\mathbf{i}_L) = \sum_{s=0}^{\nu(i_{L+1},j)-1} \delta_{k,p(i_{L+1},j,s)}\, i^{T_{p(i_{L+1},j|s+1)}(\mathbf{i}_L)}. \tag{5.4}$$

Here we have introduced the partial path

$$p(i,j\,|\,s) \overset{\text{def}}{=} (p(i,j,s),\ldots,p(i,j,\nu(i,j)-1)),$$

with the conventions

$$p(i,j\,|\,\nu(i,j)) = \emptyset, \quad T_{\emptyset}(\mathbf{i}_L) = 0.$$

Since we have previously constructed $\rho_W(i_0)$ satisfying (5.1) with $L=0$, we now have a recursive construction of $\rho_W(\mathbf{i}_L)$ satisfying the equation for all L.

We noted in Chap. 4 that if the sector code $\mathbf{i}$ is eventually periodic, i.e. has a tail $(i_1,\ldots,i_p)^\infty$, then the recursion formula (2.13) implies asymptotic temporal scaling with a scale factor $\omega_T > 1$ equal to the largest eigenvalue of $\mathbf{A}(\mathbf{i}_p)^T = (\mathbf{A}(i_1)\cdots\mathbf{A}(i_p))^T$, provided that certain minor technical assumptions on the matrices $\mathbf{A}(i)$ are valid. Eventual periodicity of $\mathbf{i}$ also leads, under appropriate circumstances, to asymptotic scaling of the global translation vectors. To see this, define a vector of return times modulo 4,

$$\boldsymbol{\tau}(\mathbf{i}_L) = (\tau_0(\mathbf{i}_L), \tau_1(\mathbf{i}_L), \ldots, \tau_{J(h(i_L))-1}(\mathbf{i}_L)), \quad \tau_j(\mathbf{i}_L) = T_j(\mathbf{i}_L) \bmod 4.$$

For sufficiently large L_0, $i_{L_0+np} = i_{L_0}$, and the vectors $\boldsymbol{\tau}(\mathbf{i}_{L_0+np})$, $n = 0,1,\ldots$ have the same length and are related by

$$\boldsymbol{\tau}(\mathbf{i}_{L_0+(n+1)p}) = \boldsymbol{\tau}(\mathbf{i}_{L_0+np}) \cdot \mathbf{A}(\mathbf{i}_p).$$

But there are at most 4^J distinct $\boldsymbol{\tau}$ vectors of length J, and so the sequence $\{\boldsymbol{\tau}(\mathbf{i}_{L_0+np}),\ n = 0,1\ldots\}$ is eventually periodic with some period N, starting with some $n = n_0$. From (5.4) we see that the sequence of matrices $M_{jk}(\mathbf{i}_L)$ becomes periodic, with period $P = Np$, starting at $L_1 = L_0 + n_0 p$,

$$M_{jk}(\mathbf{i}_{L+P}) = M_{jk}(\mathbf{i}_L),\ L \geq L_1.$$

Introducing

$$\hat{\mathbf{M}}_k = \mathbf{M}(\mathbf{i}_{L_1+k}),\ k = 1,2,\ldots,P,$$

we have the following recursion relation for the global translation vector,

$$\mathbf{d}(\mathbf{i}_L) = (d_0(\mathbf{i}_L), d_1(\mathbf{i}_L), \ldots, d_{J(h(i_L))-1}),$$

$$\mathbf{d}(\mathbf{i}_{L_1+nP}) = (\hat{\mathbf{M}}_P \cdots \hat{\mathbf{M}}_2 \cdot \hat{\mathbf{M}}_1)^n \cdot \mathbf{d}(\mathbf{i}_{L_1}),$$

and so $\mathbf{d}(\mathbf{i}_{L_1+nP})$ will scale, for asymptotically large n as the nth power of the largest-magnitude eigenvalue ω_W of the *global recursion matrix*

$$\mathbf{M} = \hat{\mathbf{M}}_P \cdots \hat{\mathbf{M}}_2 \cdot \hat{\mathbf{M}}_1,$$

that is, if $\mathbf{M} \cdot \boldsymbol{\xi} = \omega_W \boldsymbol{\xi}$, then

$$\mathbf{d}(\mathbf{i}_{L_1+nP}) \sim \omega_W^n \boldsymbol{\xi}, \tag{5.5}$$

provided that the eigenvector $\boldsymbol{\xi}$ of ω_W is unique and $\mathbf{d}(\mathbf{i}_{L_1})$ has a non-zero projection along $\boldsymbol{\xi}$. These last conditions are not automatically satisfied by the complex matrix $\mathbf{M}$, and so one has to be careful to verify them explicitly in specific models.

Invariant sectors of sawtooth kicked-oscillator models whose global translation vectors satisfy (5.5), with $\omega_W > 1$ will be called *globally expansive*. In Sect. 5.2, after developing some necessary machinery of symbolic dynamics, we will explore the implications of global expansivity for the asymptotic long-time behavior of sawtooth kicked-oscillator maps.

5.2 Long-time asymptotics

We now return to the question of the long-time behavior of globally expansive kicked-oscillator models. What we will be looking for is asymptotic power-law behavior, meaning roughly that an orbit initiated in the cell labeled by complex coordinate ζ_0 has points ζ_t which, for sufficiently large t, have $|\zeta_t| \sim At^{\mu}$, where A and μ are positive constants and corrections are of negligible relative size. We do not insist that the power-law growth be uniform in time, since there are known examples where the orbit exhibits the phenomenon of recurrence, returning sporadically to the vicinity of the origin. Thus, for $t \to \infty$, the sequence of ratios $|\zeta_t|/t^{\mu}$ have an accumulation point, not a true limit, at A. Formally this is expressed as

$$\limsup_{t\to\infty} \frac{|\zeta_t|}{t^{\mu}} = A. \tag{5.6}$$

To simplify our discussion power-law behavior, we will make certain basic assumptions. First of all, we assume that for level L larger than or equal to some fixed minimum L_*, the modulo-4 return-time vector $\boldsymbol{\tau}$ is independent of L, as are the incidence matrix $\mathbf{A}$ and the global recursion matrix $\mathbf{M}$. For an invariant sector with global expansivity, these properties can be achieved by suitably telescoping the recursive tiling structure (see Sect. 4.8). in addition, we assume that the vector of elementary lattice displacements $\mathbf{d}(\mathbf{i}_L)$ and the coefficient vectors $\mathbf{c}_j$ of (5.2) all have non-zero components in the direction of $\boldsymbol{\xi}$, the eigenvector of $\mathbf{M}$ with largest eigenvalue $\omega_W > 1$. Note that $\mathbf{c}_\mathbf{j}$, where $\mathbf{j}$ is any piece of any path $p(i,j)$, is independent of level due to our assumption on $\boldsymbol{\tau}$.

The principal benefit of the assumptions of the preceding paragraph is that we have both upper and lower bounds on the exponential growth of arbitrary lattice excursions as a function of level. Consider a piece of an orbit along a path (part $\mathbf{j}$ of some $p(i,j)$) at level $L \geq L_*$. If the orbit fragment starts at ζ, it will end at $i^{\tau(\mathbf{j})}\zeta + \Delta\zeta_L(\mathbf{j})$ where

$$\Delta\zeta_L(\mathbf{j}) = \mathbf{c}_\mathbf{j} \cdot \mathbf{d}(\mathbf{i}_L).$$

From (5.5) , the assumed non-vanishing of $\mathbf{c}_\mathbf{j} \cdot \boldsymbol{\xi}$, and the fact that there are only a finite number of possible $\mathbf{j}$, we have

$$\alpha_- \leq \frac{|\Delta\zeta(\mathbf{j})|}{\omega_W^L} \leq \alpha_+, \tag{5.7}$$

where $\alpha_\pm$ are positive bounds independent of L and $\mathbf{j}$.

We now want to use the inequalities (5.7) to obtain an upper bound on the lattice excursion from ζ_0 at time 0 to ζ_t at time t along an orbit of W. An optimal decomposition of the orbit into a concatenation of return-orbit fragments is suggested by the symbolic representation of Sect. 3.1. Suppose that the local components of the initial and final points correspond, respectively, to symbolic codes

$$\sigma = ((i_0, j_0, t_0), (i_1, j_1, t_1), \ldots, (i_L, j_L, t_L), (i_{L+1}, j_{L+1}, t_{L+1}), \ldots),$$

$$\sigma' = ((i'_0, j'_0, t'_0), (i'_1, j'_1, t'_1), \ldots, (i_L, j_L, t'_L), (i_{L+1}, j_{L+1}, t_{L+1}), \ldots),$$

where $t < t'$ and the symbols coincide from level $L+1$ onward (as they must, for some L, if the codes correspond to points on the same orbit). To get from initial to final code as efficiently as possible, we adopt the strategy of first applying W $\nu(i_0, j_0) - t_0$ times to reach (at least) level 1, then applying $\rho_W(i_0)$ $\nu(i_1, j_1) - t_1 - 1$ times to reach (at least) level 2, etc. until we reach level L, where application of $\rho_W(\mathbf{i}_{L-1})$ $t'_L - t_L - 1$ times brings the symbols into coincidence. Now we work our way down the hierarchy, applying $\rho_W(\mathbf{i}_{L-2})$ t'_{L-1} times, $\rho_W(\mathbf{i}_{L-3})$ t'_{L-2} times, until finally application of W t'_0 times completes the job.

Note that the levels $L_1, L_2, \ldots$ visited on the way up the hierarchy are not necessarily consecutive, thanks to the possible presence of maximal symbols with $t_K = \nu(i_K, j_K) - 1$. Thus, after the first $\nu(i_0, j_0) - t_0$ iterations of W, if t_1 is maximal, it will be reset to zero and t_2 incremented by one, unless it too is maximal, etc. Thus L_k, $k = 1, 2, \ldots$, is the kth non-zero level such that T_{L_k} is non-maximal. Similarly, on the way down the hierarchy, the return maps are applied only at levels M_k, $k = 1, 2, \ldots$, where t'_{M_k} is non-zero.

The sequence of codes in the above can be followed easily by keeping track of the changes in the t-components of the symbols σ_K, $k = 0, 1, \ldots, L$ (since the i-components do not change and the j-components are completely determined, reading from right to left, by the path constraints), as shown in the following diagram.

$$
\begin{array}{l}
(t_0, t_1, \ldots, t_L) \\
\searrow \\
(0, \ldots, 0, t_{L_1} + 1, t_{L_1+1}, \ldots, t_L) \\
\searrow \\
(0, \ldots, 0, t_{L_2} + 1, t_{L_2+1}, \ldots, t_L) \\
\searrow \\
\vdots \\
\searrow \\
(0, 0, \ldots, 0, t_{L+1}) \\
\downarrow \\
(0, 0, \ldots, 0, t'_L) \\
\swarrow \\
(0, \ldots, 0, t'_{M_n}, \ldots, t'_L) \\
\swarrow \\
\vdots \\
\swarrow \\
(0, \ldots, 0, t'_{M_1}, \ldots, t'_L) \\
\swarrow \\
(t'_0, t'_1, \ldots, t'_L)
\end{array}
$$

To obtain an upper bound on $|\zeta_T - \zeta_0|/\omega_W^{L(t)}$, we apply (5.7) to each of the return-map excursions with level $\geq L_*$ and sum up. The contributions from levels below L_* will be of order $\omega_W^{-L(t)}$. Thus

$$
\frac{|\zeta_t - \zeta_0|}{\omega_W^{L(t)}} \leq \alpha_+ \left(1 + \sum_{k=1}^{m} \omega_W^{L_k - L(t)} + \sum_{k=1}^{n} \omega_W^{M_k - L(t)}\right) + O(\omega_W^{-L(t)}). \tag{5.8}
$$

Since $\omega_W > 1$ and $L(t) - L_k \geq 1, L(t) - M_k \geq 1$, we have

$$
\limsup_{t \to \infty} \frac{|\zeta_t - \zeta_0|}{\omega_W^{L(t)}} \leq \alpha_+ \left(\frac{\omega_W + 1}{\omega_W - 1}\right).
$$

To show that the lim sup is non-zero, let us assume the contrary. Then infinitely many of the quantities $|\zeta_k - \zeta_0|/\omega_W^{L_{k-1}}$, where ζ_k is the starting point of the level L_k piece of the orbit, are less than $\alpha_-/2 - \varepsilon$, where ε is an arbitrarily small positive number. For such k values, we have

$$
\frac{|\zeta_{k+1} - \zeta_0|}{\omega_W^{L_k}} \geq \frac{|\Delta \zeta_k|}{\omega_W^{L_k}} - \frac{|\zeta_k - \zeta_0|}{\omega_W^{L_{k-1}}} \geq \alpha_- - (\frac{\alpha_-}{2} - \varepsilon) = \frac{\alpha_-}{2} + \varepsilon.
$$

But this would imply the existence of infinitely many orbit points ζ_k with $|\zeta_k - \zeta_0|/\omega_W^{L_{k-1}}$ more than $\alpha_-/2$, contradicting the assumption of vanishing limsup.

The temporal inequalities are easier to derive, since they involve only sums over positive quantities. In parallel with (5.8), we have

$$\beta_- \leq \limsup_{t\to\infty} \frac{t}{\omega_T^{L(t)-1}} \leq \beta_+ \left(\frac{\omega_T+1}{\omega_T-1}\right) + O(\omega_T^{-L(t)+1}).$$

Introducing

$$\delta = \frac{\log \omega_W}{\log \omega_T},$$

we have

$$\beta_+^{-\delta}\left(\frac{\omega_T-1}{\omega_T+1}\right)^{\delta} \leq \limsup_{t\to\infty} \frac{\omega_W^{L(t)-1}}{t^{\delta}} \leq \beta_-^{-\delta}. \tag{5.9}$$

Combining the limsup inequalities, we see that $\limsup_{t\to\infty} \frac{|\zeta_t - \zeta_0|}{t^{\delta}}$ is finite and non-zero. This establishes the connection between global expansivity and asymptotic power-law behavior that we were seeking.

5.3 Quadratic examples

In Lowenstein et al. (2005) the issues of global expansivity and asymptotic power laws were studied thoroughly in a number of sawtooth kicked-oscillator models with quadratic irrational parameters. The models exhibited a wide range of behaviors, including both bounded and unbounded orbits, with the latter exhibiting logarithmic, sub-diffusive, diffusive, and super-diffusive growth at infinity. The results are summarized in Table 5.1.

Table 5.1 Summary of local and global scaling parameters for the quadratic kicked-oscillator models. The last two lines refer to the residual set and discontinuity set of Model III, respectively.

λ	ω	ω_T	ω_W	μ	behavior
$\sqrt{2}$	$3-2\sqrt{2}$	9	1	0	bounded
$-\sqrt{2}$	$3-2\sqrt{2}$	9	1	0	bounded
$(-1+\sqrt{5})/2$	$(3-\sqrt{5})/2$	4	1	0	bounded
$(-1-\sqrt{5})/2$	$(3-\sqrt{5})/2$	4	1	0	bounded
$(1+\sqrt{5})/2$	$(3-\sqrt{5})/2$	4	4	1	ballistic
$(1-\sqrt{5})/2$	$(3-\sqrt{5})/2$	4	4	1	ballistic
$-\sqrt{3}$	$7-4\sqrt{3}$	25	4	.430677	sub-diffusive
$\sqrt{3}$(A)	$2-\sqrt{3}$	4	2	.5	diffusive
$\sqrt{3}$(B)	$2-\sqrt{3}$	5	2	.4306770	sub-diffusive
$\sqrt{2}$(A)	$3-2\sqrt{2}$	9	1	0	logarithmic
$\sqrt{2}$(B)	$3-2\sqrt{2}$	9	5	.732487	super-diffusive

Perhaps the most interesting of these quadratic examples is Model II of Sect. 2.3, with $\lambda = -3\tau = \sqrt{3}$, for which there are two distinct scaling sequences, A and B, each with its own scaling domain, limit point, temporal scale factor, and fractal dimension. In the case of hierarchy A, with limit point (τ, τ), local geometric scale factor $\omega_K = 2 - \lambda$ and temporal scale factor $\omega_T = 4$, there are 23 sub-domains, including 11 boundary segments. For simplicity, we limit our attention to those aperiodic orbits for which which the level-L sub-domain indices j_L, $L = 0, 1, 2, \ldots$ are restricted to the values $7, 10, 21, 22$. Relabeling these $0, 1, 2, 3$, respectively, we write down (from Kouptsov et al., 2002) the return paths and return times,

$$p(0) = (3,1,0), \quad p(1) = (3,1,1,2,0) \quad p(2) = (3,1,2,2,0) \quad p(3) = (3,2,0),$$

$$T_0(L) = T_3(L) = 26880 \times 4^L + 53, \quad T_1(L) = T_2(L) = 53760 \times 4^L - 53,$$

so that the modulo-4 return-time array is

$$\boldsymbol{\tau}(L) = (1,3,3,1),$$

independent of level L. The resulting global recursion matrix is

$$\mathbf{M} = \begin{pmatrix} 1 & i & 0 & 1 \\ 1 & 1-i & i & -1 \\ 1 & -i & 1+i & -1 \\ 1 & 0 & i & 1 \end{pmatrix},$$

with dominant eigenvalue $\omega_W^{(A)} = 2$ and corresponding eigenvector $(1,0,0,1)$. The lowest level global displacement vector,

$$\mathbf{d}(0) = (22 + 310i, 18 - 8i, 18 + 22i, -8 + 310i),$$

and for asymptotically large L,

$$\mathbf{d}(L) = \mathbf{M}^L \cdot \mathbf{d}(0) \sim 328i(\omega_W^{(A)})^L (1,0,0,1).$$

The invariant subset of orbits is indeed globally expansive. One can, moreover, verify that the coefficient vectors $c_{\mathbf{j}}$ for partial paths $\mathbf{j}$ have nonzero projections along $(1,0,0,1)$, and so the orbits should display asymptotic power-law behavior, with exponent

$$\mu_A = \frac{\log \omega_W^{(A)}}{\log \omega_T^{(A)}} = \frac{\log 2}{\log 4} = \frac{1}{2}.$$

The power law is thus a diffusive one.

As a simple example of an aperiodic orbit in this sector, we chose an initial point with a period-1 admissible symbol sequence,

$$\sigma = (1,1)^\infty, \qquad \mathbf{x}(\sigma) = \frac{1}{39}(-1422 + 821\sqrt{3},\, 7770 - 4486\sqrt{3}).$$

Here we used (3.5) to calculate $\mathbf{x}(\sigma)$ as a pair of algebraic numbers in $\mathbb{Q}(\lambda)$. A plot of the orbit (55344 iterations of $\rho(0)$) is shown in Fig. 5.2. Although it is restricted to a zero-area pseudo-chaotic web, the orbit is clearly exploring the plane in a fairly uniform manner, moving outward, on the average, at the rate of a random walk. The high level of self-similarity is due to the choice of periodic symbol sequence, and is made more evident in a log-log plot, in Fig. 5.3, of the squared distance from the origin as a function of iteration number.

For the scaling domain for scaling sequence B, the limit point is $(41 - 24\sqrt{3}, 41 - 24\sqrt{3})$, and although the local geometric scale factor is again $2 - \sqrt{3}$, the temporal scale factor is now $\omega_T^{(\mathrm{B})} = 5$. In this case, we consider all 6 polygonal sub-domains, designated 0,3,5,6,8,10 for Model II in Sect. 2.3, relabeling them as 0,1,2,3,4,5, respectively. Their paths and modulo-4 return times are, from Kouptsov et al. (2002),

$$p(0) = (3), \qquad p(1) = (3,1,4,1,2),\ p(2) = (3,1,4,5,4,1,2)$$
$$p(3) = (3,1,4,0,4,4,1,2),\ p(4) = (3,2), \qquad p(5) = (3,2,2),$$
$$\boldsymbol{\tau}(L) = (3,0,3,3,2,1).$$

Since $\boldsymbol{\tau}(L)$ is independent of L, we can calculate a level-independent global recursion matrix,

$$\mathbf{M} = \begin{pmatrix} 0 & 0 & 0 & 1 & 0 & 0 \\ 0 & 0 & 1 & i & -i & 0 \\ 0 & 1-i & 1 & 1 & -1-i & i \\ -i & 1-i & 1 & 1 & -1 & 0 \\ 0 & 0 & 1 & -i & 0 & 0 \\ 0 & 0 & 1-i & -1 & 0 & 0 \end{pmatrix},$$

with dominant eigenvalue $\omega^{(\mathrm{B})} = 2$ and corresponding eigenvector $(2i, -1+i, 4i, 4i, 2+2i, 2)$. As in the A sector, the hypotheses of global expansivity and asymptotic power-law behavior are satisfied, this time with the sub-diffusive exponent

$$\mu_{\mathrm{B}} = \frac{\log 2}{\log 5} < \frac{1}{2}.$$

In Fig. 5.4 we show the orbit of the point $x((1,2),(1,1)^\infty)$ (1438699 iterations of $\rho(0)$). Like the A orbit, this one explores large regions of the plane, moving to infinity at a rate slower than that of a normal random walk. The log-log distance vs. time plot in Fig. 5.5 shows an interesting aspect of this orbit: in spite of the steady progress toward infinity, the orbit is recurrent, i.e. it returns infinitely many times to the vicinity of the origin.

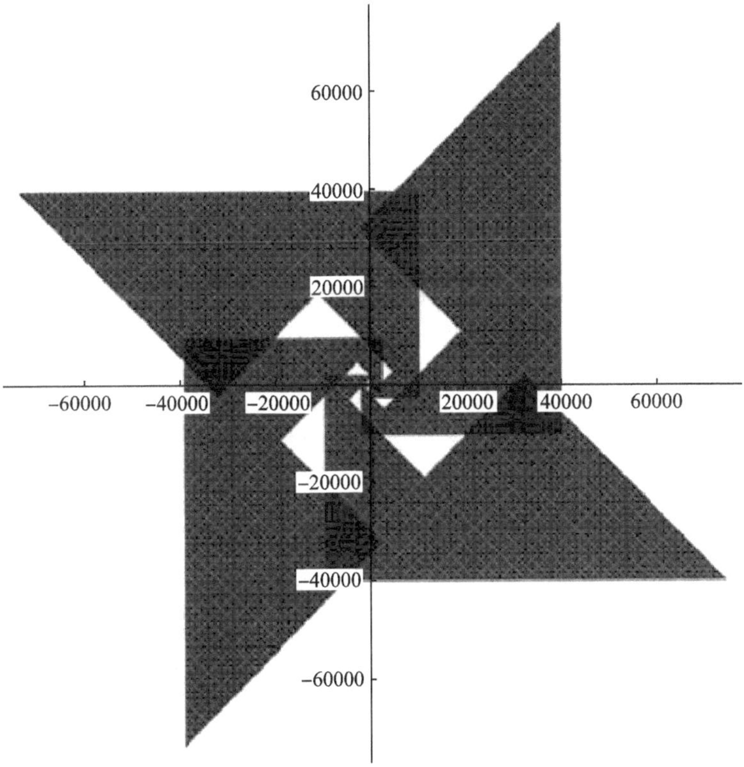

Fig. 5.2 First 10^4 $\rho_W(0)$ iterations of the aperiodic orbit starting at $\mathbf{x}(\sigma_A)$, $\sigma_A = (1,1)^\infty$.

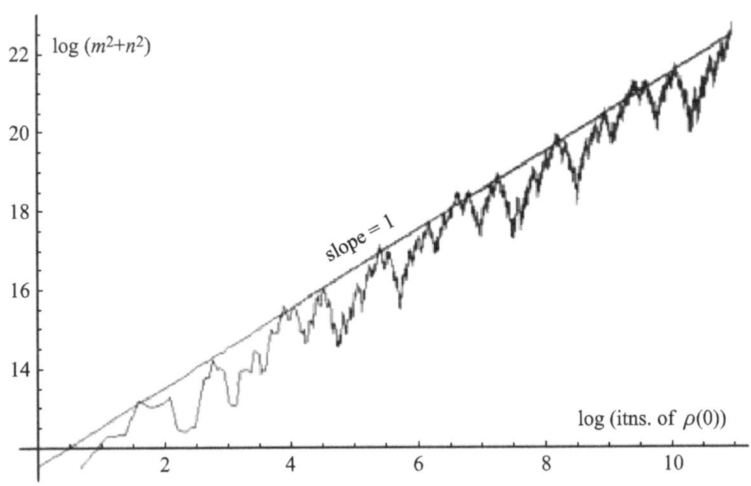

Fig. 5.3 Log-log plot of squared distance versus number of iterations of $\rho_W(0)$.

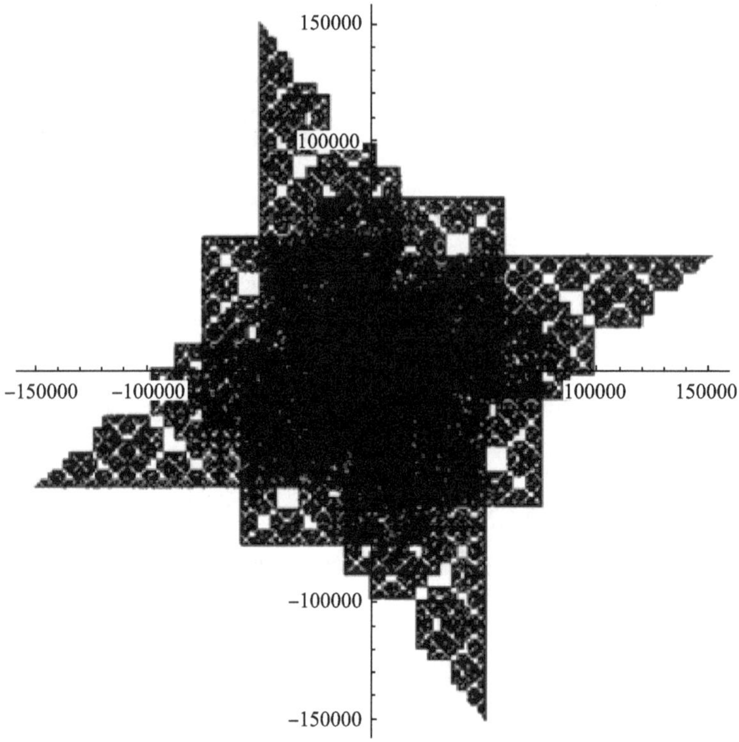

Fig. 5.4 First 1438699 $\rho_W(0)$ iterations of the aperiodic orbit starting at $\mathbf{x}(\sigma_B)$, $\sigma_B = (1,2)(1,1)^\infty$.

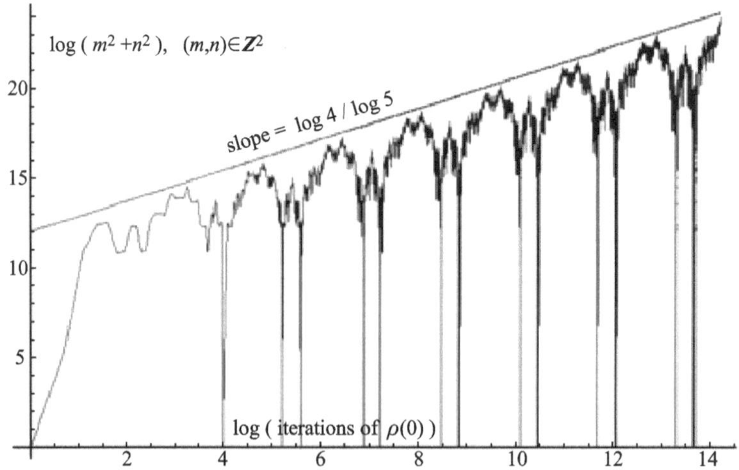

Fig. 5.5 Log-log plot of the squared distance vs. the number of iterations of $\rho_W(0)$.

5.4 Cubic examples

We now consider some more complicated, but highly instructive, examples from the $\pi/7$ model. We recall that the set Σ of discontinuity-avoiding aperiodic orbits decomposes into infinitely many invariant sectors, each labeled by a sequence of indices $\mathbf{i} = (i_0, i_1, \ldots)$. To explore the phenomena of global expansivity and asymptotic power-law behavior, we restrict ourselves to eventually periodic index sequences with period 1.

5.4.1 Orbits in the $(0, k, 6^\infty)$ *sectors*

As our first, we consider 3 sequences of the form $(0, k, 6, 6, 6, 6 \ldots)$, $k = 9, 11, 13$, where we are using the same notation as in Section 3.1.2. These sequences correspond to the three branches of the recursive tiling sub-tree shown in Fig. 5.6. The recursive localization of the base domains associated with these sequences is shown in Fig. 5.7. The embeddings $g_0, g(6), g(9), g(11), g(13)$ are given, using the notation of Sect. 3.1.2, by

$$g_0 = T_{(\lambda^{-1},0)}(-1)S_{\lambda^{-1}},$$

$$g(6) = S_{8+\lambda-3\lambda^2},$$

$$g(9) = RC^6\, T_{(-1+\lambda,1)}\, S_{4-4\lambda+\lambda^2},$$

$$g(11) = C^6\, T_{(0,-5+3\lambda)}\, S_{-9+5\lambda},$$

$$g(13) = RC^{11}\, T_{(1+\lambda-\lambda^2, 8-\lambda-2\lambda^2)}\, S_{13-4\lambda^2}.$$

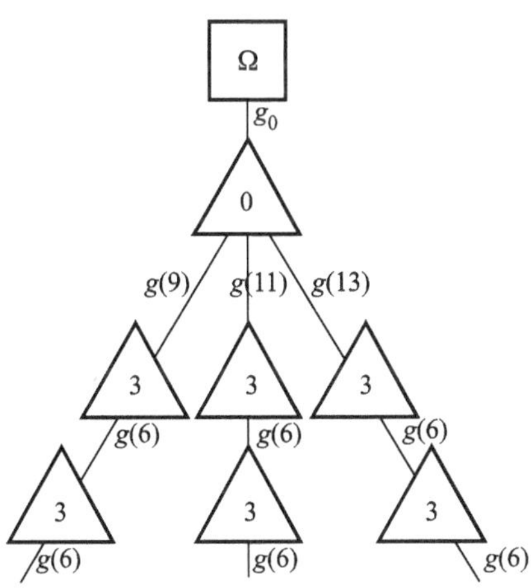

Fig. 5.6 Sub-tree of the $(0, k, 6^\infty)$ sectors.

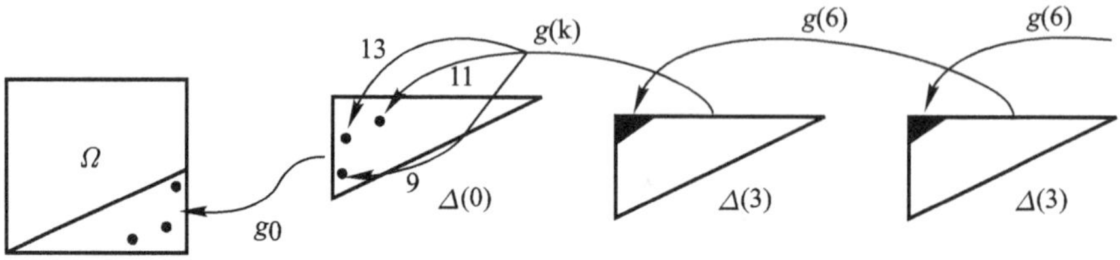

Fig. 5.7 Embedding scheme for the $(0,k,6^{\infty})$ sectors.

To explore the possibility of global expansivity in the sectors $(k,6,6,\ldots)$, we now examine the return-time vectors for increasing level. The level-0 dressed domain is a triangle equivalent to $\Delta^{[0]}$, with 4 subdomains and the return-time vector

$$\mathbf{T}(0) = (2,6,3,3) \equiv (2,2,3,3) \bmod 4.$$

The level-1 vectors are obtained by right-multiplication by the incidence matrices $\mathbf{A}(9), \mathbf{A}(11), \mathbf{A}(13)$:

$$\begin{aligned}
&\mathbf{T}(0,9) = \mathbf{T}(0) \cdot \mathbf{A}(9) = (76,226,448,214,1078,502,952,1654,1978,5056)\\
&\equiv (0,2,0,2,2,2,2,0,2,2,0) \bmod 4,\\
&\mathbf{T}(0,11) = \mathbf{T}(0) \cdot \mathbf{A}(11) = (246,810,1764,1074,5250,2466,\\
&\quad 4158,8034,8634,22062)\\
&\equiv (2,2,0,2,2,2,2,2,2,2,2) \bmod 4,\\
&\mathbf{T}(0,13) = \mathbf{T}(0) \cdot \mathbf{A}(13) = (174,576,1260,732,3612,1692,\\
&\quad 2898,5532,6024,15402)\\
&\equiv (2,0,0,0,0,0,2,0,0,2) \bmod 4.
\end{aligned}$$

We note that already at the first level, a simplification has occurred: the entries of the modulo-4 return-time vectors are all even. This is due to a peculiarity of the incidence matrices $\mathbf{A}(9), \mathbf{A}(11), \mathbf{A}(13)$, namely that their third and fourth rows are identical, combined with the fact that the third and fourth components of $\mathbf{T}(0)$ also coincide. Even though these circumstances will not repeat themselves at higher levels, the restriction to even entries cannot be undone, since multiplication of an even integer by any integer is necessarily even. A similar "ratchet" effect will occur if, at some level, the entries are all zeros.

For levels greater than 1, we calculate

$$\begin{aligned}
&\mathbf{T}(0,9,6,6,...,6) = \mathbf{T}(0,9) \cdot \mathbf{A}(6) \cdot A(6) \cdots A(6) \equiv (0,2,0,2,2,2,0,2,2,0) \bmod 4,\\
&\mathbf{T}(0,11,6,6,...,6) = \mathbf{T}(0,11) \cdot \mathbf{A}(6) \cdot A(6) \cdots A(6)\\
&\equiv (0,0,0,0,0,0,0,0,0,0) \bmod 4,\\
&\mathbf{T}(0,13,6,6,...,6) = \mathbf{T}(0,13) \cdot \mathbf{A}(6) \cdot A(6) \cdots A(6)\\
&\equiv (0,2,0,2,2,2,0,2,2,0) \bmod 4,
\end{aligned}$$

We see that for $L \geq 2$, the modulo-4 return-time vectors are independent of L, and so we can employ (5.4) to obtain the complex global recursion matrices:

$$\mathbf{M}(0,9,6,6,...,6) = \mathbf{M}(13,6,6,...,6) = \begin{pmatrix} 1 & 0 & 0 & 0 & 0 & 0 & 1 & 0 & 0 & 0 \\ 0 & 0 & 0 & 0 & 0 & 0 & -1 & 0 & 1 & 0 \\ 2 & 0 & 0 & 0 & 0 & 0 & 1 & 0 & 0 & 1 \\ 0 & 0 & 0 & 0 & 0 & 1 & -1 & 0 & 0 & 0 \\ 0 & 0 & 0 & 0 & 0 & 4 & -1 & 0 & -3 & 0 \\ 0 & 0 & 0 & 0 & 0 & 2 & -1 & 0 & -1 & 0 \\ 1 & 0 & 0 & 0 & 0 & 0 & 1 & 0 & 0 & 0 \\ 0 & 0 & 0 & 0 & 0 & 6 & -1 & 0 & -5 & 0 \\ 0 & 0 & 0 & 0 & 0 & 0 & -1 & 0 & 1 & 0 \\ 1 & 0 & 0 & 0 & 0 & 0 & 1 & 0 & 0 & 0 \end{pmatrix},$$

$$\mathbf{M}(0,11,6,6,\ldots,6) = A(6)^T$$

The matrix $\mathbf{M}(0,9,6,6,...,6)$ departs slightly from the standard scenario in that its dominant eigenvalue, $\omega_W = 2$ has a 2-dimensional eigenspace. This is not an essential problem, however, and we find that the complex displacement vector $\mathbf{d}(0,9,6,6,\ldots,6)$ behaves, for $L \to \infty$, as 2^L times a unique eigenvector of $\mathbf{M}(0,9,6,6,...,6)$,

$$\mathbf{d}(0,9,6,6,\ldots,6) \sim 2^L(1,-1,2,-1,-1,-1,1,-1,-1,1).$$

Similarly,

$$\mathbf{d}(0,13,6,6,\ldots,6) \sim 2^L \times \frac{3}{2}(-1,1,-2,5,17,9,-1,25,1,-1).$$

Since $\mathbf{A}(6)^T$ has maximum eigenvalue

$$\omega_T = 8 + 2\sqrt{7} = 13.2915\ldots,$$

the sectors $(0,9,6,6,\ldots)$ and $(0,13,6,6,\ldots)$ are globally expanding, and we can expect asymptotic behavior with lattice distance increasing as t^{μ}, with

$$\mu = \frac{\log \omega_W}{\log \omega_T} = \frac{\log 2}{\log \omega_T} = 0.267922\ldots.$$

Since μ is smaller than the value $1/2$ characterizing a random walk (normal diffusion), we label the asymptotic behavior as sub-diffusive.

In the sector $(0,11,6,6,\ldots)$, the displacement vectors are proportional asymptotically to ω_T^L times an eigenvector of $\mathbf{A}(6)$ with all positive components. The lattice distance increases proportional to the time, analogous to ballistic motion.

5.4.2 Numerical investigations

Now that we have a theoretical understanding of the power-law asymptotic behavior of the aperiodic orbits in the sectors $(0,k,6,6,\ldots)$, it is worthwhile to see how the concepts are realized in detail for specific initial conditions. We choose two admissible codes σ_A and σ_B in the sectors $(0,9,6^\infty)$ and $(0,11,6^\infty)$, respectively, calculating the corresponding points in Ω using (3.5),

$$\sigma_A = (0,1,0)(9,8,0)\,(6,8,8)^\infty, \quad \sigma_B = (0,1,0)(11,8,0)\,(6,8,8)^\infty,$$

$$\begin{aligned}
\mathbf{u}_A &= g_0\, g(9)\,(1 - C^{10} S_{8+\lambda-3\lambda^2})^{-1} \rho^{[3]8}(\mathbf{0}) \\
&= \frac{1}{71}\left(864 + 391\lambda - 471\lambda^2, 347 - 237\lambda + 30\lambda^2\right), \\
\mathbf{u}_B &= g_0\, g(11)\,(1 - C^{10} S_{8+\lambda-3\lambda^2})^{-1} \rho^{[3]8}(\mathbf{0}) \\
&= \frac{1}{71}\left(279 + 624\lambda - 425\lambda^2, -2087 - 1078\lambda + 1241\lambda^2\right),
\end{aligned}$$

In our calculations, we treat λ as an abstract symbol, using the relation $\lambda^3 = \lambda^2 + 2\lambda - 1$ to simplify polynomial expressions with exact rational coefficients. In this way the dynamical map can be iterated arbitrarily many times without being contaminated by round-off error. Floating point arithmetic enters only at the stage of plotting the results.

Our first experiment was aimed at visualizing the local orbits. In Figs. 5.8 and 5.9 we see the result of iterating the local map K 10,000 times. The two local orbits are not qualitatively very different. Both show evidence of island-around-island structures reminiscent of sticky orbits found in chaotic models. To see this more clearly, we zoomed in to a small square of side 0.0013 the width of the fundamental cell, increasing the number of iterations to 10^7 to obtain high resolution. In Fig. 5.10 the island-around-island structure is quite evident. Not pictured are the islands themselves; these are periodic heptagons of various sizes which in fact occupy 100 percent of the area, with the aperiodic orbits confined to a set of measure zero.

Now we want to see the W orbits of the same initial points on the plane. From our analysis based on global expansivity, we expect the orbits to behave quite differently for long times: orbit A has scaling exponent 0.27, and hence should proceed relatively slowly to infinity (much slower than a random walk, with exponent 1/2), while orbit B, perhaps after a transient period, should move rapidly outward, with distance from the origin increasing proportional to the time. We begin by calculating the first 10,000 points on each orbit (these are the same points whose local coordinates we have already seen).

The resulting pictures, in Fig. 5.11, are dramatically different: the points of orbit A fill out a cloud of approximate radius 3 surrounding the origin, with little discernable structure. Orbit B, on the other hand, has already become an "accelerator mode", spiraling outward at an approximately constant rate in narrow strips surrounding the x and y axes.

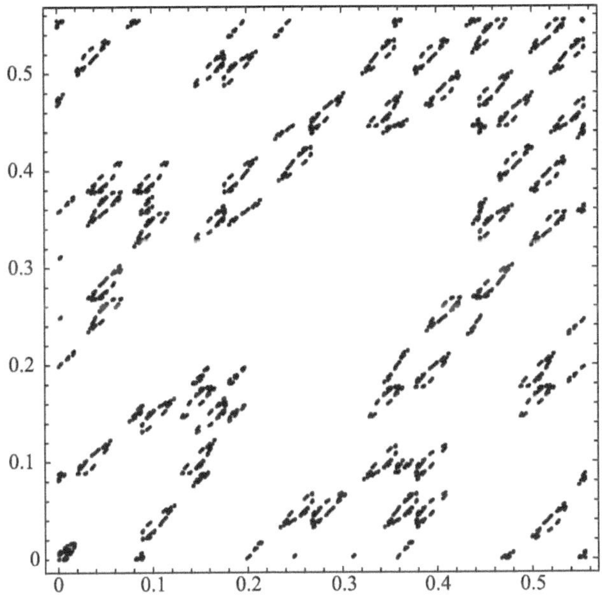

Fig. 5.8 Local orbit (10^4 iterations) of $\mathbf{x}(\sigma_A)$.

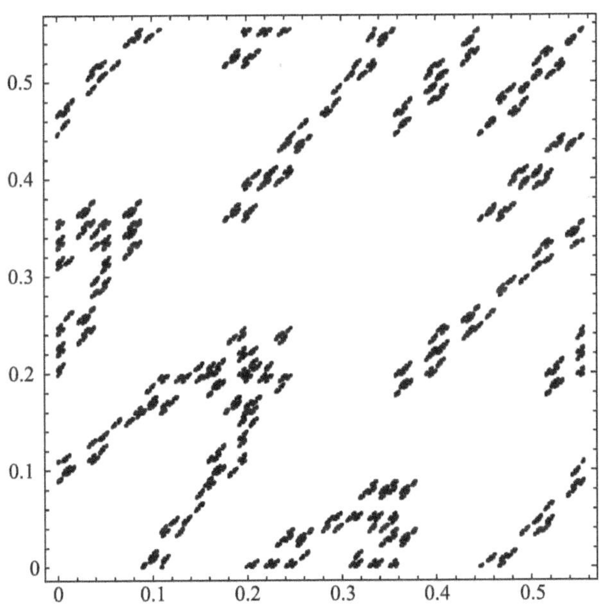

Fig. 5.9 Local orbit (10^4 iterations) of $\mathbf{x}(\sigma_B)$.

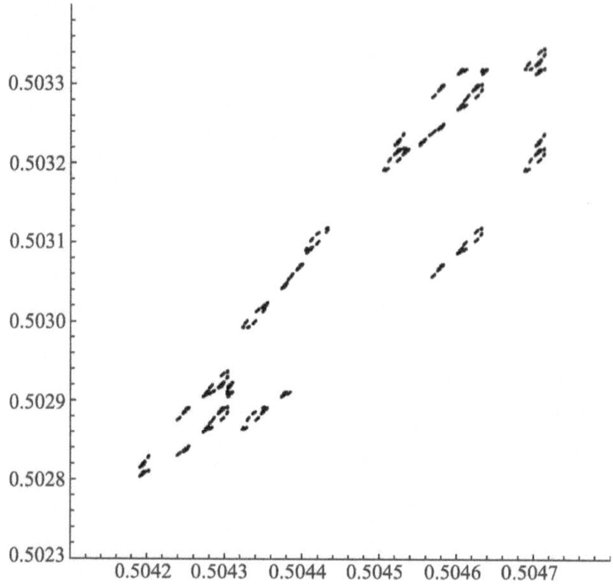

Fig. 5.10 Detail of the $\mathbf{x}(\sigma_B)$ orbit, showing island-around-island structure. Actual islands are periodic heptagons, not shown.

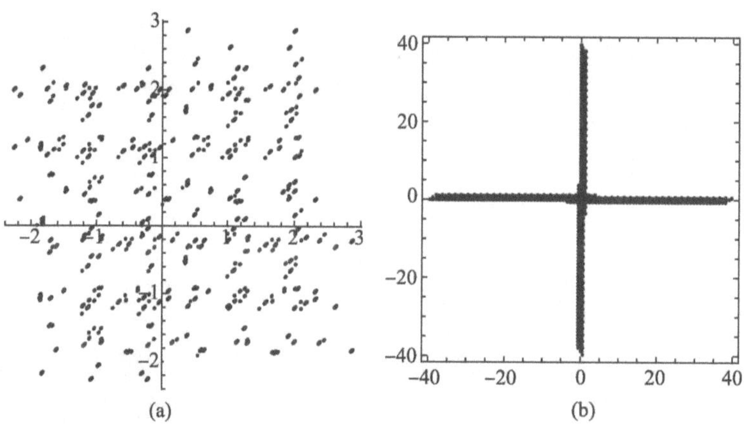

Fig. 5.11 Global orbits (10^4 iterations) of $x(\sigma_A)$ and $x(\sigma_B)$.

A clearer picture of what is going on is obtained by increasing the number of iterations to 1 million, as shown in Fig. 5.12. Orbit A has continued to make its way slowly outward, and some interesting geometric patterns have emerged. With a 0.27 power law, we would expect a 100-fold increase in t to produce an increase of linear size of approximately $100^{0.27} \approx 3.5$, which is consistent with the figure. Orbit B, on the other hand, has continued its linear progress to infinity, with the maximum distance increasing by a factor of approximately 100. Moreover, the departure from the axes is now below the resolution of the graphics.

Finally, in Fig. 5.13 we plot the distance from the origin as a function of time, on a log-log scale, for the two orbits. In the sub-diffusive case, we note the apparently self-similar pattern of shifts from one level to another, including frequent returns to the neighborhood of the origin (recurrences), with the maximum level slowly increasing in accordance with the power law.

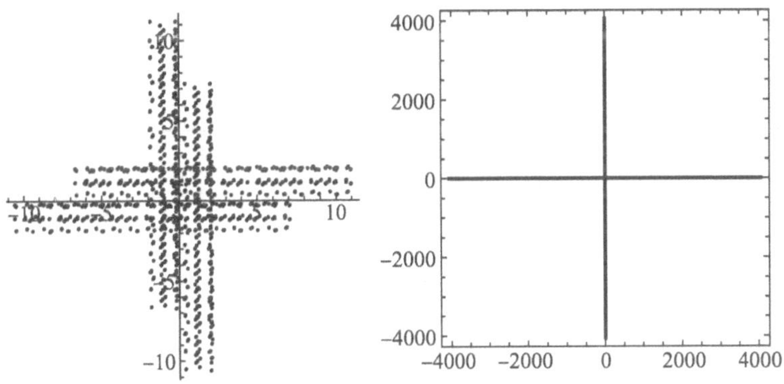

(a) (b)

Fig. 5.12 Same orbits as in Fig. 5.11, with 10^6 iterations.

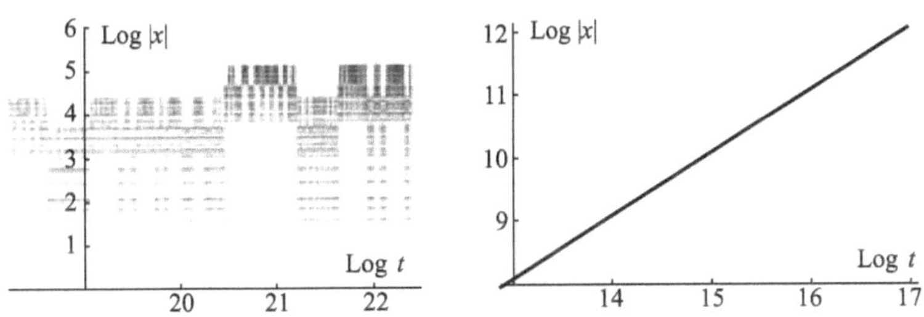

(a) (b)

Fig. 5.13 Log-log plot of distance vs. number of iterations for the orbits of **(a)** $x(\sigma_A)$ and **(b)** $x(\sigma_B)$.

5.4.3 A non-expansive sector

Not all invariant sectors of the $\pi/7$ kicked oscillator are globally expansive. It is instructive to see how this can come about, even when the global recursion matrix has an eigenvalue greater than 1 in magnitude. Our example has sector code $\mathbf{i} = (0, 10, 1^\infty)$. As in the previous cases, the modulo-4 return-time arrays are built up by repeated application of the incidence matrices:

$$\begin{aligned}
\boldsymbol{\tau}(0) &= (2,2,3,3),\\
\boldsymbol{\tau}(0,10) &= \boldsymbol{\tau}(0)\cdot \mathbf{A}(10) \bmod 4 = (2,2,2,2,2,2,2,2,2,2,2,2),\\
\boldsymbol{\tau}(0,10,1) &= \boldsymbol{\tau}(0,10)\cdot \mathbf{A}(1) \bmod 4 = (0,2,0,0,0,0,0,0,0,0,0,0)\\
\boldsymbol{\tau}(0,10,1,1) &= \boldsymbol{\tau}(0,10,1)\cdot \mathbf{A}(1) \bmod 4 = (2,2,2,2,2,2,2,2,2,2,2,2),\\
\boldsymbol{\tau}(0,10,1,1,1) &= \boldsymbol{\tau}(0,10,1,1)\cdot \mathbf{A}(1) \bmod 4 = (0,2,0,0,0,0,0,0,0,0,0,0)
\end{aligned}$$

The global recursion matrix is

$$\mathbf{M} = \mathbf{M}^{(2)}\cdot \mathbf{M}^{(1)},$$

$$M^{(l)}_{j,k} = c_{k,p(1,j)} = \sum_{t=0}^{v(1,j)-1} \delta_{k,p(1,j|t)}\, i^{\tau^{(l)}_{p(1,j|t+1)}},\ l = 1,2,$$

where

$$\boldsymbol{\tau}^{(1)} = (2,2,2,2,2,2,2,2,2,2,2,2),\quad \boldsymbol{\tau}^{(2)} = (0,2,0,0,0,0,0,0,0,0,0,0).$$

Diagonalization of $\mathbf{M}$ reveals eigenvalues $3, 1, -1$, and 0, and so one might expect a sub-diffusive asymptotic power law, with exponent

$$\mu = \frac{\log 3}{2\log\alpha} = 0.151270..., \tag{5.10}$$

where $\alpha = 19+4\sqrt{22}$ is the maximum eigenvalue of $\mathbf{A}(1)^T$. On the other hand, explicit calculation of the complex displacement vectors for increasing level gives

$$\begin{aligned}
\mathbf{d}(0) &= (1-i,-i,2,1+i),\\
\mathbf{d}(0,10) &= \mathbf{M}(0,10)\cdot \mathbf{d}(0) = (1-4i,1-4i,1-4i,\dots,1-4i),\\
\mathbf{d}(0,10,1) &= \mathbf{M}^{(1)}\cdot \mathbf{d}(0,10) = (0,1-4i,0,0,0,0,0,0,0,0,0,0),\\
\mathbf{d}(0,10,1,1) &= \mathbf{M}^{(2)}\cdot \mathbf{d}(0,10,1) = (1-4i,1-4i,1-4i,\dots,1-4i),\\
\mathbf{d}(0,10,1,1,1) &= \mathbf{M}^{(1)}\cdot \mathbf{d}(0,10,1,1) = (0,1-4i,0,0,0,0,0,0,0,0,0,0)
\end{aligned}$$

and we see that there is no global expansivity. In fact, there is no contradiction: the displacement vector $(1-4i,\dots,1-4i)$ is indeed an eigenvector of $\mathbf{M}$, but corresponds to the nonleading eigenvalue 1 instead of the dominant eigenvalue 3.

To see what happens to a typical orbit in this sector, we select the simplest admissible code and calculate the corresponding point in Ω (this parallels exactly the calculations of the previous examples, and so we omit the details:

$$\sigma_C = (0,1,0),(10,7,0),(1,7,0)^{\infty},$$

$$\mathbf{u}_C = \frac{1}{58}(86+52\lambda-47\lambda^2,-179+57\lambda+26\lambda^2).$$

The first 100,000 points of the local orbit are shown in Fig. 5.14(a), while the corresponding points of the global orbit are shown in Fig 5.15(a). Further iteration does not expand the extent of the latter at all.

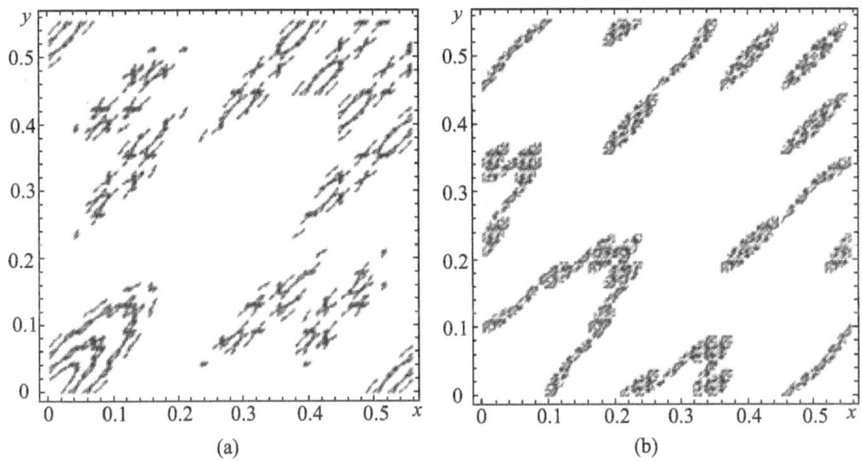

Fig. 5.14 Local orbits in sectors (**a**) $(0,10,1^\infty)$ and (**b**) $(0,12,1^\infty)$.

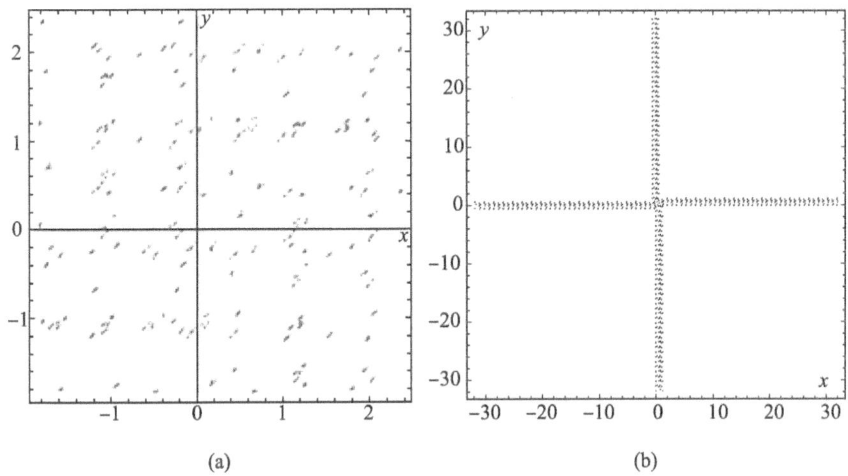

Fig. 5.15 Global orbits in sectors (**a**) $(0,10,1^\infty)$ and (**b**) $(0,12,1^\infty)$.

In contrast to this non-expansive case is the sector $(0,12,1^\infty)$, which has the same modulo-4 return time arrays and global recursion matrix as $(0,10,1^\infty)$. This time, however, the projection of the displacement vector along the dominant eigendirection is non-zero, and so there is a global scaling factor $\omega_W = 3$ and a subdiffusive power law with exponent μ given by (5.10). The local and global orbits are plotted in Figs. 5.14(b) and 5.15(b).

5.4.4 Generic behavior

In the preceding subsection, we have gotten a glimpse of the various possibilities for power-law asymptotics in the countably many invariant sectors with eventually periodic sector codes. Such behavior is not generic, of course, since the full set of sectors is uncountable. In this final subsection, we consider the generic case, for which the sector code is a random sequence of i_L compatible with the constraint $n(i_{L+1}) = h(i_L)$. Apart from the initial choice of i_0 (see Fig. 2.15), this is a ternary process: given i_L, the choice of i_{L+1} is among exactly 3 equally weighted possibilities.

First of all, let us calculate the probability distribution for the various possible modulo-4 return-time arrays $\tau(\mathbf{i}_L)$, in the limit $L \to \infty$. We suspect that, due to the 'ratchet effect' (irreversibility of the transition to $(0,0,\ldots,0)$), the probabilities may become skewed toward arrays containing only zeros. By explicit calculation, we find that from $L = 2$ onward, only the following arrays are represented:

$h(i_L) = 1$

$$(0,0,0,0,0,0,0,0,0,0,0,0),\ (0,0,2,0,0,2,0,0,0,2,2),\ (0,2,0,0,0,0,0,0,0,0,0),$$
$$(0,2,2,0,0,2,0,0,0,2,2),\ (2,2,0,2,2,0,2,2,2,0,0),\ (2,2,2,2,2,2,2,2,2,2,2),$$

$h(i_L) = 2$

$$(0,0,0,0,0,0,0,0,0,0,0,0,0),(0,0,0,0,2,0,2,0,0,0,2,0),$$
$$(0,2,0,0,0,0,0,0,0,0,0,0),$$
$$(0,2,0,0,2,0,2,0,0,0,2,0),(2,0,2,2,0,2,0,2,0,2,0,2),$$
$$(2,0,2,2,2,2,2,2,0,2,2,2),$$

$h(i_L) = 3$

$$(0,0,0,0,0,0,0,0,0,0),\quad (0,2,0,2,2,2,0,2,2,0,),$$

Multiplying (from the right) each of the above by the three appropriate incidence matrices ($\mathbf{A}(0),\mathbf{A}(1),\mathbf{A}(2)$ for $h(i_L) = 1$, $\mathbf{A}(3),\mathbf{A}(4),\mathbf{A}(5)$ for $h(i_L) = 2$, and $\mathbf{A}(6),\mathbf{A}(7),\mathbf{A}(8)$ for $h(I_L) = 3$) we find that the above set of 14 arrays is mapped into itself, with a transition matrix

$$\mathbf{R} = \begin{pmatrix}
2 & 0 & 0 & 0 & 0 & 0 & 1 & 0 & 0 & 0 & 0 & 0 & 0 & 0 \\
0 & 1 & 0 & 1 & 0 & 0 & 0 & 1 & 0 & 0 & 0 & 0 & 0 & 0 \\
1 & 0 & 0 & 0 & 0 & 1 & 1 & 0 & 0 & 0 & 0 & 0 & 0 & 0 \\
0 & 0 & 0 & 1 & 1 & 0 & 0 & 1 & 0 & 0 & 0 & 0 & 0 & 0 \\
0 & 1 & 0 & 1 & 0 & 0 & 0 & 0 & 0 & 0 & 0 & 1 & 0 & 0 \\
0 & 0 & 2 & 0 & 0 & 0 & 0 & 0 & 0 & 0 & 1 & 0 & 0 & 0 \\
2 & 0 & 0 & 0 & 0 & 0 & 1 & 0 & 0 & 0 & 0 & 0 & 0 & 0 \\
0 & 1 & 0 & 1 & 0 & 0 & 0 & 1 & 0 & 0 & 0 & 0 & 0 & 0 \\
1 & 1 & 0 & 0 & 0 & 0 & 0 & 0 & 1 & 0 & 0 & 0 & 0 & 0 \\
0 & 1 & 1 & 0 & 0 & 0 & 0 & 0 & 0 & 1 & 0 & 0 & 0 & 0 \\
0 & 0 & 1 & 0 & 1 & 0 & 0 & 0 & 0 & 1 & 0 & 0 & 0 & 0 \\
0 & 1 & 0 & 0 & 0 & 1 & 0 & 0 & 1 & 0 & 0 & 0 & 0 & 0 \\
2 & 0 & 0 & 0 & 0 & 0 & 0 & 0 & 0 & 0 & 0 & 0 & 1 & 0 \\
2 & 0 & 0 & 0 & 0 & 0 & 0 & 0 & 0 & 0 & 0 & 0 & 0 & 1
\end{pmatrix}.$$

The level-L populations of the 14 species, $n(L)$, is then given by

$$\mathbf{n}(L) = (n_1(L), n_2(L), \ldots, n_{14}(L)) = n(2) \cdot R^{L-2},$$

where, by explicit calculation,

$$\mathbf{n}(2) = (6,0,4,0,0,2,1,0,0,0,2,0).$$

The $L \to \infty$ behavior is obtained by diagonalizing the transpose of R. The largest eigenvalue is 3, with eigenvector $(2,0,0,0,0,0,1,0,0,0,0,0,0,0)$ corresponding to only arrays containing zeros. The convergence rate is determined by the second-largest eigenvalue, $2.95263\ldots$: the probability for $\tau(i_L) \neq 0$ converges to zero proportional to $(2.95263/3)^L = 0.98421^L$. The constant of proportionality is found to be $0.109446\ldots$.

We conclude that for a randomly selected invariant sector, the modulo-4 return-time array has only zeros from some level onward, with probability one. Scaling is generic, with the power-law exponent equal to unity ('ballistic' case). Moreover, the dense set of points with eventually periodic codes is drastically affected by the same mechanism: of all the points whose codes become periodic starting at level 10 or higher, at least 90% have eventually all-zero modulo-4 return-time arrays, and hence have orbits with $\mu = 1$ power laws.

References

Kouptsov K.L., Lowenstein J.H., and Vivaldi F. (2002) Quadratic rational rotations of the torus and dual lattice maps, *Nonlinearity* **15**, 1795–1842.

Lowenstein J.H., Poggiaspalla G., and Vivaldi F. (2005) Sticky orbits in a kicked-oscillator model, *Dynamical Systems* **20**, 413–451.

Chapter 6
Transport

In our discussions so far concerning the asymptotic long-time behavior of pseudochaotic kicked oscillators, we have concentrated on individual orbits, especially those within the web of discontinuity-avoiding aperiodic orbits. In the present chapter we turn our attention to the statistical properties of various classes of orbits for asymptotically long times. Once again, the recursive tiling property serves as a key organizing principle.

In defining a coarse-grained probability distribution $P((m,n),t)$, it is important to take account of the $\pi/2$ rotation incorporated in the map W. The orbits of interest actually spiral out to infinity rather than following a succession of small steps on the lattice. Obviously, our probability distribution refers to lattice *displacements* $(m_2 - m_1\tau, n_2 - n_1\tau) : m_1, m_2, n_1, n_2 \in \mathbb{Z}$ and time *differences* $t_2 - t_1$, $t_1, t_2 \in \mathbb{Z}$, with the latter always assumed to be multiples of 4 (we are not interested in the position-dependent displacements due to the $\pi/2$ rotations). Due to the invariance of the 4th-iterate map W^4 (see (1.6)), the probability distribution will then be independent of the choice of initial lattice point, allowing us to assume, for convenience, that all relevant orbits are initiated in the fundamental domain Ω.

With these considerations in mind, we adopt the following definition.

Definition 6.1. We define $P((m,n),t)$ as the probability that a W orbit initiated at time 0 at a randomly selected point in Ω will be found in the domain $\Omega + (m\tau, n\tau)$ after t iterations, t an integer multiple of 4.

We note that $P((m,n),t)$ is always positive on a finite number of lattice sites (vanishing elsewhere), and satisfies the normalization condition, for all t,

$$\sum_{m,n=-\infty}^{\infty} P((m,n),t) = 1.$$

If $\chi(\mathbf{r})$, $\mathbf{r} = (m,n) \in \mathbb{Z}^2$ is any real-valued lattice function, we define the *expectation value* of $\chi(\mathbf{r})$ at time t,

$$\langle \chi(\mathbf{r}) \rangle_t = \sum_{\mathbf{r}} P(\mathbf{r},t)\chi(\mathbf{r}). \tag{6.1}$$

Moments of the probability distribution correspond to $\chi((m,n))$ being a monomial in $m, n, |m|, |n|, |m| + |n|$, or $\sqrt{m^2+n^2}$.

In specific examples, the words "randomly selected" in the definition of $P((m, n),t)$ need to be interpreted in terms of some measure, which, as pointed out in the Introduction, is not necessarily Lebesgue measure on the plane. In fact, some of the most interesting long-time behavior is found in subsets of the exceptional set, which for all of the models of interest is of Lebesgue measure zero. For these subsets, Hausdorff measure, which we discussed in some detail in Chap. 4, serves as the basis for our probability calculations.

6.1 Probability calculation using recursive tiling

The definition of the probability distribution $P(\mathbf{r},t)$ given above suggests a direct method of calculation, namely, with the help of a random number generator, to select large numbers of initial conditions in Ω and then apply the map W iteratively on each one, recording the lattice coordinates of each orbit at a succession of time intervals. The number of visits to each lattice site could then be accumulated and eventually normalized to give the desired probability. Such a direct approach is unfortunately highly inefficient, due to the extreme degree of nonuniformity of the local phase space. Most of the computational effort would end up iterating relatively uninteresting low-period orbits which never get very far from the origin.

A more elegant strategy would make use of the time-t dynamical partition of Ω. The latter consists of a set of disjoint convex polygons, with distinct t-iteration itineraries, whose union is Ω. By construction, a polygon D in the partition can be subjected to t iterations of W without landing on a discontinuity line, ending up in the domain $\Omega + (m,n)\tau$. We then have an exact calculation of $P((m,n),t)$, namely

$$P((m,n),t) = \sum_{D \in \mathscr{D}_t(m,n)} \frac{\mathscr{A}(D)}{\mathscr{A}(\Omega)}, \tag{6.2}$$

where

$$\mathscr{D}_t(m,n) = \{D : W^t(D) \subset \Omega + (m,n)\tau\}.$$

While using the dynamical partition removes problems of convergence (for every t, the calculation is exact) and eliminates overemphasis on parts of phase space where branching has ceased, it still does not take advantage of the exponential shrinking of the time scale which renormalizability offers. For that we turn once again to the hierarchy of recursive tilings.

For simplicity, let us restrict ourselves to a model with quadratic irrational parameter and a single scaling sequence. Then, at level L, the square Ω is partitioned into the orbits of the periodic domains $\Pi_k(L')$, $L' < L$, together with those of the scaling subdomains $D_j(L)$, $j = 0,1,\ldots,J$. As usual, $D(L) \subset D_{j_0}(L-1)$ for $L \geq 1$ and some j_0, and $D(0) \subset \Omega_{k_0}$ for some k_0. We call the partition $\mathscr{P}(L)$.

Let $t_*(L)$ be the largest number of iterations such that for all $D \in \mathscr{P}(L)$, $W^t(D)$ does not intersect a discontinuity line for $t = 1,2,\ldots,t_*(L)$. Then we can use formula (6.2) to calculate $P(\mathbf{r},t)$ exactly for any $t \le t_*(L)$. Our hope is, of course, that for large L, the non-branching interval $t_*(L)$ will scale as ω_T^L. This will give an exponential shrinking of the time scale for calculations of $P(\mathbf{r},t)$, just as we had for individual orbits in Chapter 5.

At first sight, the above strategy appears to be futile. Consider any one of the level-L local return orbits $K^s(D_j(L))$, $s = 0,1,\ldots,T_j(L)$. The first $T_j(L)$ members of this sequence are members of $\mathscr{P}(L)$, but the last is not and may in fact straddle a boundary between some $D_{j_1}(L)$ and $D_{j_2}(L)$. Since it takes only one iteration of K to pass from $K^{T_j(L)-1}(D_j(L))$ to $K^{T_j(L)}(D_j(L))$, it would seem that we are stuck with $t_*(L) = 1$! Fortunately, there is a loophole in this reasoning. The fact that the adjacent domains $D_{j_1}(L)$ and $D_{j_2}(L)$ initiate return orbits with distinct level-$(L-1)$ itineraries does not imply that they will separate immediately under application of K. In fact, since both are subdomains of $D(L)$, hence of $D_{j_0}(L-1)$, they will remain glued together for at least $T_{j_0}(L-1)$ iterations. In fact, once they get mapped around the full return orbit of $D_{j_0}(L-1)$, they land in $D_{j_0}(L-2)$, and so the free ride continues for at least another $T_{j_0}(L-2)$ iterations. Eventually, the forward images of $D_{j_1}(L)$ and $D_{j_2}(L)$, still glued together, end up in Ω_{k_0}, and there is no further guarantee against landing on a discontinuity line of K. We conclude that

$$t_*(L) \ge T_*(L) \stackrel{\text{def}}{=} \sum_{L'=0}^{L-1} T_{j_0}(L'). \tag{6.3}$$

The desired large-L scaling behavior, i.e.

$$T_*(L) \sim \omega_T^L,\ L \to \infty. \tag{6.4}$$

is immediate.

Now we are in a position to write an explicit formula for $P(\mathbf{r},t)$ and its moments $\langle \chi(\mathbf{r}) \rangle_t$. Given t, we choose the partition $\mathscr{P}(L_*(t))$, where $L_*(t)$ is given by

$$L_*(t) = \min\{L : t \le T_*(L)\}. \tag{6.5}$$

For any function χ on $\mathbb{Z}^2$ and sequence of integer pairs

$$Z = (z_0, z_1, \ldots, z_{|Z|-1}), \qquad z_i \in \mathbb{Z}^2,$$

we define, for nonnegative integer $t \le T$,

$$\langle\langle \chi \rangle\rangle_t^Z = \sum_{i=0}^{|Z|-T-1} \chi(z_{i+t} - z_i). \tag{6.6}$$

Denote by $Z_D(L,j)$ (resp. $Z_\Pi(L,j)$)) the lattice coordinates of the W orbit of the centroid of $D_j(L)$ (resp. $\Pi_j(L)$), extended to a total of $T_j(L) + T_*(L)$ (resp. $t_j(L) + T_*(L)$) iterations. If $\mathscr{A}(D)$ is the area of a domain D, we define

$$\langle \chi \rangle_t = \sum_{-1<L'<L_*(t)} \sum_{j'} \frac{\mathscr{A}(\Pi_{j'}(L'))}{\mathscr{A}(\Omega)} \langle\langle \chi \rangle\rangle_t^{Z_\Pi(L',j')} + \sum_j \frac{\mathscr{A}(D_j(L_*(t)))}{\mathscr{A}(\Omega)} \langle\langle \chi \rangle\rangle_t^{Z_D(L_*(t),j)}. \tag{6.7}$$

Here level -1 is included in the first sum to include all periodic orbits which do not intersect the scaling domain $D(0)$. The same formula can be used to calculate the probability distribution $P(\mathbf{r},t)$, since, by definition (6.1),

$$P((m,n),t) = \langle \delta_m \delta_n \rangle_t, \qquad m,n,t \in \mathbb{Z}, \qquad t \geq 0, \tag{6.8}$$

where

$$\delta_n(n') \stackrel{\text{def}}{=} \begin{cases} 1 & n = n', \\ 0 & n \neq n'. \end{cases}$$

6.2 Ballistic transport in Model I

Perhaps the simplest of our examples with nontrivial global expansion is Model I, with golden mean parameter $\lambda = \frac{1}{2}(\sqrt{5}+1)$. As a warm-up calculation to illustrate the method, let us calculate the second moment $\langle m^2 + n^2 \rangle_t$for $t = 8$. As required, t is a multiple of 4. For this value of t, we need to use the tiling of (at least) level $L_*(10) = 1$ (see (6.5), noting that $T_*(1) = T_2(0) = 10$). The tiling, shown in Fig. 6.1, partitions Ω into three K-orbits of periodic pentagons, namely $\Pi_4(0), \Pi_5(0)$,

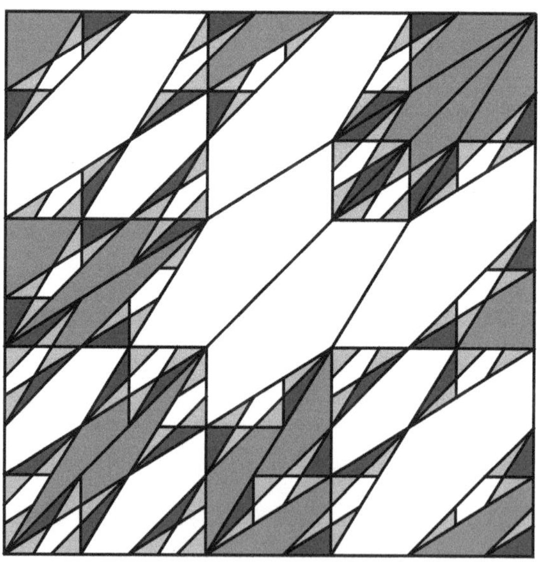

Fig. 6.1 Level-1 tiling of Ω. The periodic pentagons are colored white, while the domains $D_0(1), D_2(1)$, and $D_4(1)$ are colored gray, dark gray, and light gray, respectively.

and $\Pi_0(-1)$, complemented by the three return orbits of the three scaling subdomains $D_0(1)$, $D_2(1)$, and $D_4(1)$. Note that we can safely omit orbits of domains of dimension less than 2, since these do not contribute to the probability distribution. For the pentagons, the periods are $24, 6$, and 4, respectively, while for the subdomains of $D(1)$, the return times are, respectively, $16, 40$, and 88. The reader can easily locate all of these orbits on the figure.

As a prelude to the moment calculation, we calculate all six of the corresponding global orbits on $\mathbb{Z}^2$, remembering to extend each one by $T_*(1) = 10$ iterations. To simplify the calculation, we map pointwise the centroid of each convex polygon, since all points in the polygon have the same lattice itinerary. We get, for example,:

$$\begin{aligned} Z_\Pi(0,5) = \{&(0,0),(0,1),(1,-1),(-1,-1),(-1,1),(1,1),(1,-2),(-2,0),(0,1),\\ &(1,0),(0,-1),(-1,0),(0,0),(0,1),(1,-1),(-1,-1),(-1,1)\}. \end{aligned}$$

For the second moment at $t = 8, T_*(1) = 10$, (6.6) becomes

$$\langle\langle \chi \rangle\rangle_8^{Z_D(L,j)} = \sum_{i=0}^{T_j(L)-1} ((m_{i+8} - m_i)^2 + (n_{i+8} - n_i)^2),$$

with an analogous expression for $\langle\langle \chi \rangle\rangle_8^{Z_\Pi(L,j)}$. Note that the number of terms in the sum is precisely the return time (or period) of the orbit, since we get a single contribution from each tile in the partition of Ω. The six double-brackets are easily calculated:

$$\langle\langle \chi \rangle\rangle_8^{Z_\Pi(0,4)} = 68, \qquad \langle\langle \chi \rangle\rangle_8^{Z_\Pi(0,5)} = 8, \qquad \langle\langle \chi \rangle\rangle_8^{Z_\Pi(-1,7)} = 16,$$

$$\langle\langle \chi \rangle\rangle_8^{Z_D(1,0)} = 36, \qquad \langle\langle \chi \rangle\rangle_8^{Z_D(1,2)} = 104, \qquad \langle\langle \chi \rangle\rangle_8^{Z_D(1,4)} = 240.$$

To proceed, we need the weights in the probability sum, given by the areas

$$\begin{aligned} \mathscr{A}(\Pi_4(0)) &= -(199 + 123\tau)/2, & \mathscr{A}(\Pi_5(0)) &= -(11 + 7\tau)/2,\\ \mathscr{A}(\Pi_7(-1)) &= -9\tau - 29/2, & \mathscr{A}(D_0(1)) &= -13\tau - 21,\\ \mathscr{A}(D_2(1)) &= -17\tau - 55/2, & \mathscr{A}(D_4(1)) &= (89 + 55\tau)/2, \end{aligned}$$

as well as

$$\mathscr{A}(\Omega)^{-1} = \tau^{-2} = 2 + \tau.$$

Putting everything together, we get, from (6.7),

$$\langle x^2 + y^2 \rangle_8 = 54 + 32\tau = 2.22291\ldots$$

The probability distribution at time $t = 8$ can be determined from (6.8). The result is displayed in Fig. 6.2, with the shorthand notation

$$a = 233/2 + 72\tau, \quad b = 161 + 199\tau/2, \quad c = 89/2 + 55\tau/2, \quad d = 34 + 21\tau,$$

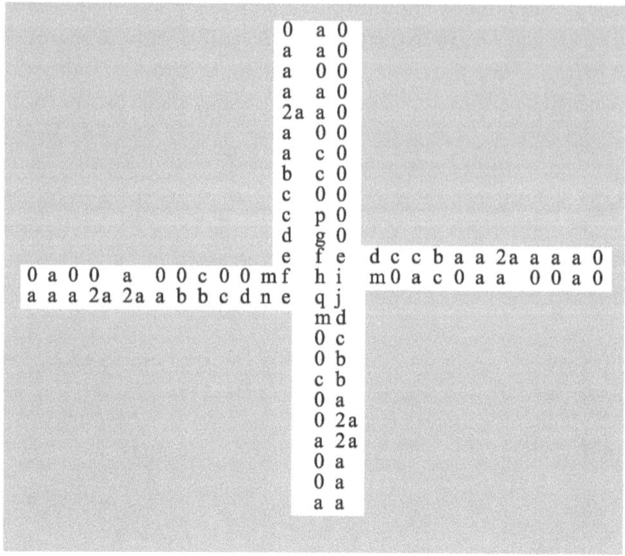

Fig. 6.2 Probability distribution after 8 iterations of W. The symbols a–h and m–q stand for the values listed in (6.9). The probability is zero over the entire gray region.

$$
\begin{aligned}
e&=-571/2-353\tau/2, \quad f=-359-691\tau/2, \quad g=-1063/2-657\tau/2, \qquad (6.9)\\
h&=-2807/2-1735\tau/2, \quad m=-271/2-335\tau/2, \quad n=-453/2-140\tau,\\
p&=-21/2-13\tau/2, \quad q=-521-322\tau.
\end{aligned}
$$

We now are ready for a more ambitious application of formula (6.7), namely the calculation of the moments $\langle m^2+n^2\rangle_t$ and $\langle m^4\rangle_t$ of $P((m,n),t)$ over the entire range of t from 0 to $218450 = T_*(8)$, using the level-8 partition of Ω. The latter is comprised of the 20 return orbits listed, together with their respective areas and return times, in Table 6.1. For the sake of efficiency, the first step of our procedure is to calculate and store the lattice coordinates of the various extended return orbits. The moments are then calculated systematically using (6.7). If an exact result for a specific time interval is desired, it is essential to employ arbitrary-precision integer arithmetic and computer algebra. For example, as a test case, we calculated the second moment for $t = 218448$, the largest multiple of 4 less than or equal to $T_*(8)$, obtaining

$$
\langle m^2+n^2\rangle_{218448} = -22775702889924663042 78-14076158503644190279 30\,\tau
$$

Log-log plots for 500 t values, $932 \le t \le 218448$, are displayed in Fig. 6.3 on a logarithmic time scale. The chosen $\log_{10} t$ values are evenly spaced, up to round-off. The t^ν increase of $\langle (m^2+n^2)^{\frac{\nu}{2}}\rangle_t$ comes as no surprise, given that all but one of the 20 lattice orbits expands outward ballistically. The exception is the tightly bound orbit of $\Pi_5(0)$, which has negligible effect on the moments.

Table 6.1 Areas and return times for the 20 domains participating in the calculation of moments for $t \leq 218448$.

Domain	Area	Return Time
$D_0(8)$	$-14930352-9227465\tau$	218456
$D_2(8)$	$(-39088169-24157817\tau)/2$	655360
$D_4(8)$	$31622993+39088169\tau/2$	1529168
$\Pi_4(7)$	$70711162-87403803\tau/2$	436904
$\Pi_5(7)$	$-70711162-87403803\tau/2$	54616
$\Pi_4(6)$	$(-20633239-12752043\tau)/2$	109224
$\Pi_5(6)$	$(-1149851-710647\tau)/2$	13656
$\Pi_4(5)$	$-3010349/2-930249\tau$	27304
$\Pi_5(5)$	$-167761/2-51841\tau$	3416
$\Pi_4(4)$	$-219602-271443\tau/2$	6824
$\Pi_5(4)$	$-219602-271443\tau/2$	856
$\Pi_4(3)$	$(-64079-39603\tau)/2$	1704
$\Pi_5(3)$	$(-3571-2207\tau)/2$	216
$\Pi_4(2)$	$-9349/2-2889\tau$	424
$\Pi_5(2)$	$-521/2-161\tau$	56
$\Pi_4(1)$	$-682-843\tau/2$	104
$\Pi_5(1)$	$-38-47\tau/2$	16
$\Pi_4(0)$	$(-199-123\tau)/2$	24
$\Pi_5(0)$	$(-11-7\tau)/2$	6
$\Pi_7(0)$	$-29/2-9\tau$	4

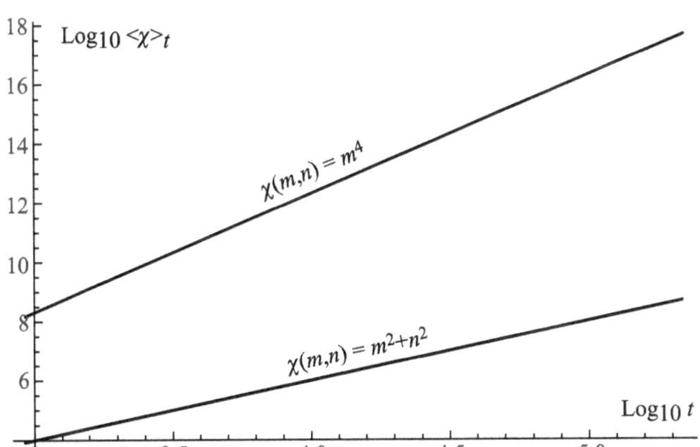

Fig. 6.3 Log-log plots of $\langle m^2+n^2\rangle_t$ and $\langle m^4\rangle_t$.

6.3 Subdiffusive transport in Model II

We now turn to a statistical description of the long-time behavior on the pseudochaotic web (residual set) of a globally expansive model with quadratic irrational parameter. Our choice is Model II, with $\lambda = -\tau^{-1} = \sqrt{3}$, where we have already found asymptotic diffusive and subdiffusive asymptotic behavior for orbits in the parts of the web associated with the scaling sequences A (limit point (τ, τ)) and B (limit point $(41 + 72\tau, 41 + 72\tau)$), respectively. In the present section we will focus on the latter, making use of Hausdorff measure to assign weights to the orbits contributing to the probability distribution $P(\mathbf{r}, t)$.

Apart from the restricted measure, concentrated on the pseudochaotic web of Case B, the strategy for calculating the expectation values $\langle \chi \rangle_t$ is the same as that used in the preceding sections. In fact, it is simpler: only the second summation (over level-$L_*(t)$ tiles) is relevant, since the periodic domains do not contribute to the measure:

$$\langle \chi \rangle_t = \sum_j \mu_H(D_j(L_*(t))) \langle\langle \chi \rangle\rangle_t^{Z_D(L_*(t),j)}, \tag{6.10}$$

where μ_H is normalized Hausdorff measure (concentrated on the Case B web, with dimension $d_H = -\log 5 / \log \omega$). According to Sec. 4.4, the measure assigned to the level-zero atoms is given by

$$\mu_H(D_0(0)) = \mu_H(D_{10}(0)) = \frac{3}{c}, \qquad \mu_H(D_3(0)) = \frac{20}{c},$$

$$\mu_H(D_5(0)) = \mu_H(D_6(0)) = \frac{15}{c}, \qquad \mu_H(D_8(0)) = \frac{19}{c},$$

$$c = 5298000.$$

A serious practical problem which we must face in the present example is the fact that the return times are extremely large, even for the lowest levels of the scaling sequence. Fortunately we can significantly improve on the mapping method of the previous example by exploiting the fact that the lattice itineraries are entirely determined by the symbolic dynamics, i.e. Vershik updating of code sequences, with the least significant code symbol at each step determining the current lattice displacement. Only elementary operations with integers are involved (no symbolic manipulation of algebraic numbers), and so the calculations can be performed rapidly using a short program in a standard compiled computer language.

Let us carry out the above program in a numerical experiment which probes several levels of the scaling hierarchy. We calculate $\langle \chi \rangle_t$ for

$$\chi = r^k, \; k = 1, 2, \; r = |x| + |y|,$$

for a sample of time intervals $t_k, k = 1, \ldots, N$, all multiples of 4, approximately equally spaced on a logarithmic scale, with

$$1052 \leq t_k \leq t_{\max} = 72000000.$$

To do this, we calculate symbolically the orbits of $D_j(5)$, $j = 0,3,5,6,8,10$ for $T_j(5)+t_{\max}$ iterations, with

$$\{T_j(5) : j = 0,3,5,6,8,10\} =$$
$$\{59107675, 236507588, 295690483, 295690483, 118215350, 177323025\},$$

using a codometer with 7 registers,

$$\left(\begin{pmatrix}0\\j_0\\t_0\end{pmatrix}\begin{pmatrix}1\\j_1\\t_1\end{pmatrix}\begin{pmatrix}1\\j_2\\t_2\end{pmatrix}\begin{pmatrix}1\\j_3\\t_3\end{pmatrix}\begin{pmatrix}1\\j_4\\t_4\end{pmatrix}\begin{pmatrix}1\\j_5\\t_5\end{pmatrix}\begin{pmatrix}1\\j_6\\t_6\end{pmatrix}\right).$$

The paths $(p(1,j,t) : t = 0,\ldots,\nu_j)$ assigned to all but the zeroth register are

$$(6),\quad (6,3,8,3,5),\quad (6,3,8,10,8,3,5),\quad (6,3,8,0,8,8,3,5),\quad (6,5),\quad (6,5,5).$$

The lengths of the paths $(p(0,j,t) : t = 0,\ldots,\nu_j)$ assigned to register 0 are given by

$$\nu_0(0) = 19459, \qquad \nu_0(3) = 76580, \qquad \nu_0(5) = 93115,$$

$$\nu_0(6) = 93115, \qquad \nu_0(8) = 38918, \qquad \nu_0(10) = 58377.$$

For the sake of efficiency, the corresponding paths (sequences of 0's,1's and 2's) were worked out and stored in preparation for the calculation of moments.

According to the prescription (6.10), the exact moment calculation involves measurements on only six orbits, namely those of $D_j(5), j = 0,3,5,6,8,10$. In the initial codometer settings for these, we set $j_5 = j, t_5 = 0$, which requires $(j_k,t_k) = (6,0)$ for $k = 0,1,2,3,4$. The last register is then set to a value compatible with $j_5 = j$, in such a way that the full $T_j(5)+t_{\max}$ iterations can be accommodated.

Before proceding to the calculation of moments, let us briefly examine the six orbits on which all the measurements are to be made. Figure 6.4 provides a low-resolution (2000×2000 pixels) view of the footprints of these orbits, revealing the asymptotic large-scale self-similarity. Log-log plots of the expectation values $\langle\chi\rangle_t$ for $\chi = |m|+|n|$ and $(|m|+|n|)^2$ are shown in Fig.6.5. The average slopes are consistent with the power laws found in Sect. 5.3, namely

$$\langle(|m|+|n|)^k\rangle_t \sim t^{ks}\, G_k(t)$$

where

$$s = \frac{\log 2}{\log 5} = 0.430676558\ldots$$

and, as we see in Fig. 6.6, the form factors G_k are apparently bounded period-1 functions of $\log_5 t$.

Using essentially the same program for $\chi = \delta_m\,\delta_n$, we have also calculated the probability distribution $P((m,n),t)$ for a single value of t, namely 72000000. The result is displayed with 2000×2000 resolution in Fig. 6.7. Here we have used colors to indicate the magnitude of the probability on a logarithmic scale.

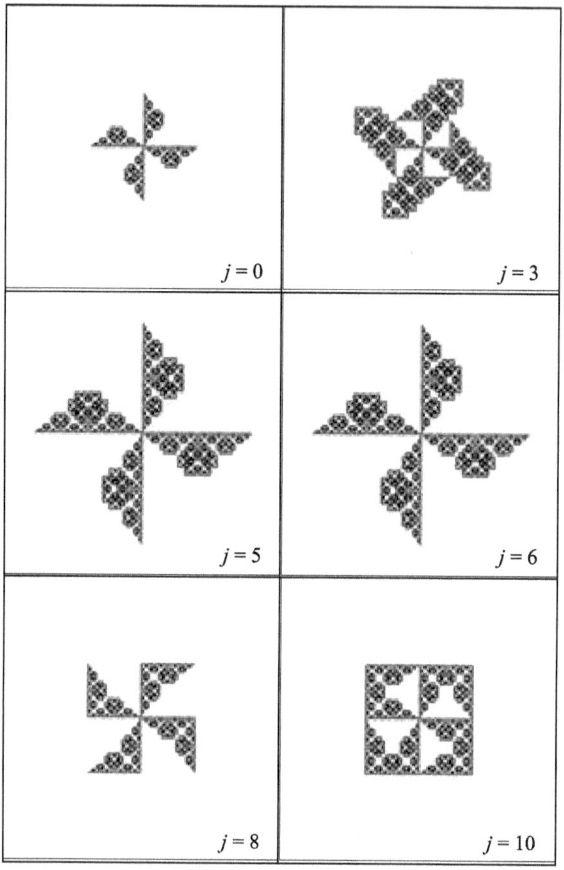

Fig. 6.4 Footprints of the 6 orbits contributing to the calculation of moments.

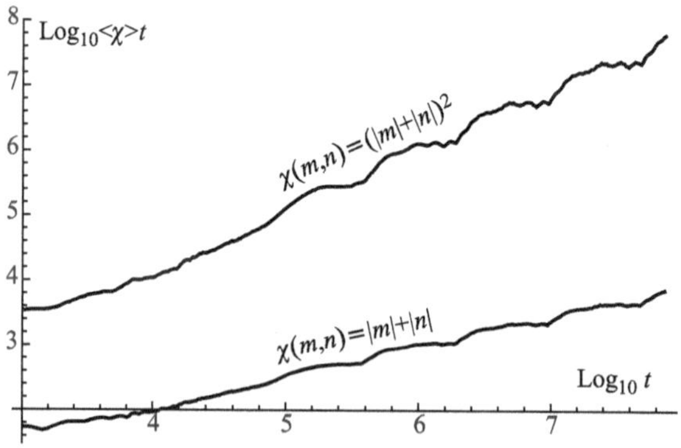

Fig. 6.5 Log-log plots of the expectation values $\langle \chi \rangle_t$ for $\chi = |m| + |n|$ and $(|m| + |n|)^2$.

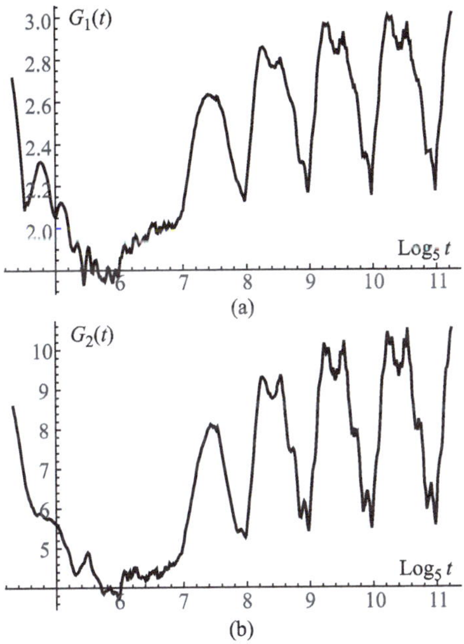

Fig. 6.6 Form factors of the expectation values $\langle \chi \rangle_t$ for $\chi = |m| + |n|$ and $(|m| + |n|)^2$. They appear to be tending toward log-periodicity.

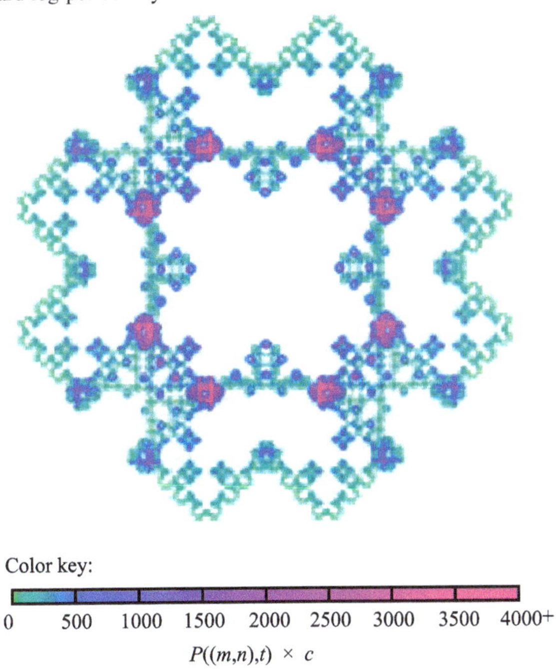

Fig. 6.7 Probability distribution $P((m,n),t)$, multiplied by the normalization factor $c = 5298000$, for web B, $-1000 \leq m,n < 1000, t = 72000000$. (See the color figure at the end of the book.)

6.4 Diffusive transport in Model II

Following the same pattern as in the preceding section, we now study the statistical long-time behavior of the other pseudochaotic web for $\lambda = \sqrt{3}$, namely the one whose scaling sequences tend to $(\tau,\tau) = (-1/\sqrt{3}, -1/\sqrt{3})$. Once again, formula (6.10) is applicable, this time with Hausdorff dimension $d_H = -\log 4/\log\omega \approx 1.05$, and level-0 measures

$$\mu_H(D_7(0)) = \mu_H(D_{10}(0)) = \mu_H(D_{21}(0)) = \mu_H(D_{22}(0)) = \frac{1}{161280},$$

and $\mu_H(D_j(0)) = 0$ for $j \neq 7, 10, 21, 22$.

Our numerical calculations of the probability distribution and its moments parallels that of the preceding section. Again we probe the recursive tiling hierarchy up to level 5, but because of the slower growth relate of the scaling sequence, the return times $T_j(5)$ and critical time $t_*(5)$ are smaller. Our sample of 401 time intervals extends from $t = 524$ through $t_{\max} = 9000000$. Again we generate the lattice orbits via a 7-register codometer, with paths $p(1,j)$ for nonzero registers given by

$$\begin{aligned} p(1,7) &= (22,10,7), \\ p(1,10) &= (22,10,10,21,7), \\ p(1,21) &= (22,10,21,21,7), \\ p(1,22) &= (22,21,7). \end{aligned}$$

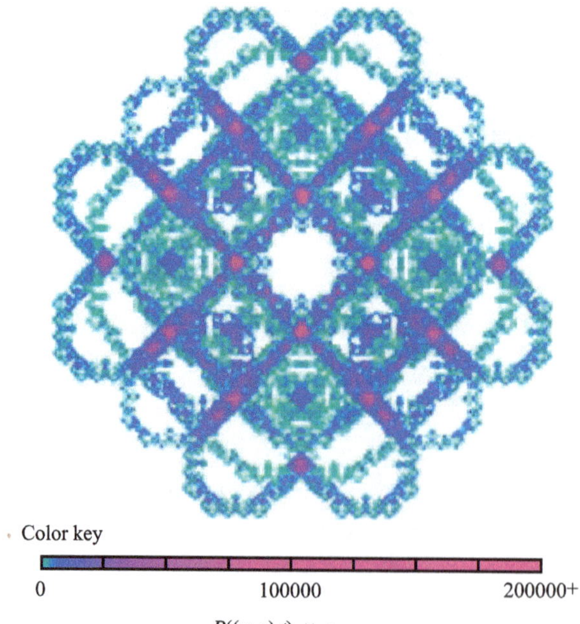

Fig. 6.8 Probability distribution $P((m,n),t)$, multiplied by the normalization factor $c = 161,280$, for web A, $-550 \leq m,n < 550, t = 9000000$. (See the color figure at the end of the book.)

The paths used by the zeroth register, of lengths

$$\nu_7(0) = \nu_{22}(0) = 26933, \qquad \nu_{10}(0) = \nu_{21}(0) = 53707,$$

are calculated by iterating the local map and storing the resulting code sequences in advance of the moment calculations.

Using (6.10), we have calculated the expectation values $\langle (|m| + |n|)^k \rangle_t$, $k = 1, 2$, as well as the $P(\mathbf{r}, t)$ for $t = 9000000$. This involves symbolic updating over the return orbits of $D_j(5)$, $j = 7, 10, 21, 22$, with return times

$$T_7(5) = T_{22}(5) = 27525173, \qquad T_{10}(5) = T_{21}(5) = 55050187.$$

The results are displayed in Figs. 6.8 and 6.9.

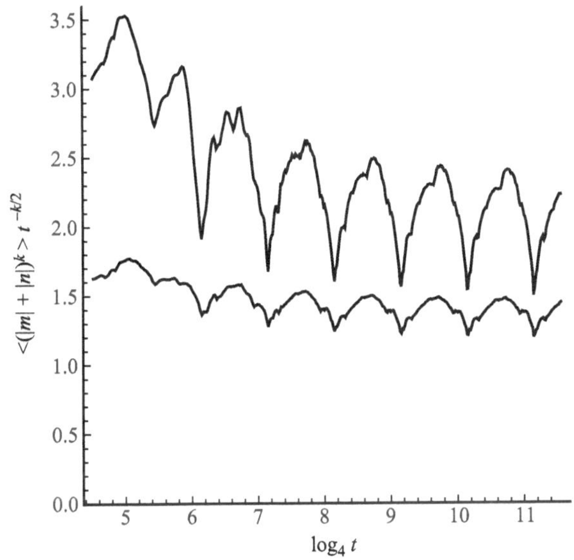

Fig. 6.9 Form factors of the expectation values $\langle \chi \rangle_t$ for $\chi = |m| + |n|$ and $(|m| + |n|)^2$ in web A, plotted versus $\log_4 t$. They appear to be tending toward periodicity.

6.5 Superdiffusive transport in Model III

As we see in Table 5.1, the long-time asymptotics of Model III is globally expansive, but with a dramatic difference between the aperiodic orbits of the residual set and those residing on the $x = 0$ discontinuity line. Among the latter, one finds (Lowenstein et al., 2005) asymptotic superdiffusive behavior for orbits which pass through a Cantor set $\mathscr{C}$ of dimension $-\log 5 / \log(3 - 2\sqrt{2})$. In the present section we study the statistical properties of these orbits, including numerical calculations of several moments of a suitable probability distribution.

Note that we are dealing here with a set of measure zero, not only with respect to 2-dimensional Lebesgue measure on the plane, but also with respect to one-dimensional Lebesgue measure on the discontinuity set. Not unexpectedly, we will use Hausdorff measure on $\mathscr{C}$, as constructed in Sect. 4.5, in order to formulate an acceptable probability measure for calculating expectation values $\langle \chi \rangle_t$.

Suppose we are interested in calculating $\langle \chi \rangle_t$ for a given t. Following Sect. 6.1, we calculate $L = L_*(t)$ from (6.5) and partition $\mathscr{C}$, apart from boundary points, into subsets covered by forward images of the one-dimensional atoms $D_j(L)$, $j = 1,3,5,7$. A recursive prescription for this covering is given in Sect. 2.5.2 and summarized in Sect. A.3.7 of the Appendix. Specifically, the covering tiles are precisely those encountered along the respective level-L return orbits of the 4 atoms. The number $N_j(L)$ of each tile type was found in Sect. 2.5.2 to be

$$N_1(L) = N_7(L) = 6 \cdot 5^L, \qquad N_3(L) = N_5(L) = 9 \cdot 5^L,$$

and their Hausdorff measures, from Sect. 4.5 , are

$$\mu_H(D_1(L)) = \mu_H(D_5(L)) = \frac{1}{20}, \qquad \mu_H(D_3(L)) = \mu_H(D_7(L)) = \frac{1}{60}.$$

Suppose we have calculated all $6 \times 5^{L+1}$ baseline segments $\mathscr{D}_\alpha$ covering $\mathscr{C}$ at level $L = L_*(t)$, choosing from each a representative point $\mathbf{u}_k$ with algebraic number coordinates. It is important to note that the point need not be an element of $\mathscr{C}$, since all orbits initiated in the covering segment have the same lattice itinerary for times no longer than t. For simplicity, we could choose $\mathbf{u}_k$ to be the midpoint of $\mathscr{D}_k$. Denoting by $[\mathbf{u}]$ the integer lattice coordinates of the planar point $\mathbf{u}$, we have for the χ moment for time interval t the formula

$$\langle \chi \rangle_t = \sum_{\alpha=1}^{6\times 5^{L+1}} \mu_H(\mathscr{D}_\alpha)\chi([W^t \mathbf{u}_\alpha]). \tag{6.11}$$

Since every $\mathscr{D}_\alpha$ is isometric to one of the $D_j(L)$, $j = 1,3,5,7$, we can rewrite (6.11) as

$$\langle \chi \rangle_t = \sum_{j=1,3,5,7} \mu_H(D_j(L)) \sum_{\mathscr{D}_\alpha \cong D_j(L)} \chi([W^t \mathbf{u}_\alpha]). \tag{6.12}$$

Because of lattice translation invariance of the dynamics, we can replace the second summation in (6.12) by an expression which involves operations on a single orbit originating in the base tile $D_j(L)$:

$$\langle \chi \rangle_t = \sum_{j=1,3,5,7} \mu_H(D_j(L)) \sum_{\substack{k=0 \\ W^k \mathbf{u}_j \in \mathscr{B}}}^{T_j(L)} \chi([W^{k+t} \mathbf{u}_j] - [W^k \mathbf{u}_j]). \tag{6.13}$$

As in the example of Sect. 6.3, the exact iteration of the planar map W using symbolic manipulation of algebraic numbers can be circumvented by making use of the fact that the motion on the integer lattice is completely driven by the symbolic dynamics. We have adopted this strategy in the following numerical experiment.

Our goal is to calculate $\langle \chi \rangle_t$ for

$$\chi = r^k,\ k = 1,2,4,6,\ r = |x| + |y|,$$

for a sample of time intervals t_k, $k = 1, \ldots, N$, all multiples of 4, approximately equally spaced on a logarithmic scale, with

$$1000 \leq t_k \leq t_{\max} = 16000000.$$

To do this, we calculate symbolically the orbits of $D_j(7)$, $j = 1,3,5,7$ for $T_j(7) + t_{\max}$ iterations, with

$$\{T_j(7) : j = 1,3,5,7\} = 171874384, 315207204, 358253925, 214921105,$$

using a telescoped codometer with 7 registers,

$$\left(\begin{pmatrix} i_2 \\ j_2 \\ t_2 \end{pmatrix} \begin{pmatrix} i_3 \\ j_3 \\ t_3 \end{pmatrix} \begin{pmatrix} i_4 \\ j_4 \\ t_4 \end{pmatrix} \begin{pmatrix} i_5 \\ j_5 \\ t_5 \end{pmatrix} \begin{pmatrix} i_6 \\ j_6 \\ t_6 \end{pmatrix} \begin{pmatrix} i_7 \\ j_7 \\ t_7 \end{pmatrix} \begin{pmatrix} i_8 \\ j_8 \\ t_8 \end{pmatrix} \right).$$

Here the indices i_L take on the successive values $0,1,2,1,2,1,2$. The first register is a telescoping of levels $-1,0,1$, and 2, with $0 \leq t_2 < T_{j_2}(2)$, in contrast to the other registers where, as usual, $0 \leq t_L < v_{j_L}$.

The instantaneous state of the system consists of a pair $(V, \mathbf{r})$ where V is a codometer state and $\mathbf{r} = (m,n)$ is a point on the lattice $\mathbb{Z}^2$. Under the dynamical map W,

$$(V, \mathbf{r}) \mapsto (V', \mathbf{r}'),$$

where V' is obtained from V using the rules of Vershik updating (see Sect. 3.2), and

$$\mathbf{r}' = (m', n') = (n, -m) + d(p(0, j_2(V), t_2(V))).$$

Here the path function gives the current atom index (corresponding to one of the 12 generating domains Ω_j), and $d(j)$ is the corresponding lattice displacement, namely

$$\begin{gathered} d(1) = d(3) = d(5) = d(6) = d(7) = d(8) = (0,0); \\ d(2) = d(4) = (0,-1); \end{gathered}$$

$$d(0) = d(10) = d(9) = d(11) = (0,1);$$

For each of the intimal domains $D_j(7)$, we start with an initial state

$$\psi = \left(\left(\begin{pmatrix} 0 \\ 8 \\ 0 \end{pmatrix} \begin{pmatrix} 1 \\ 8 \\ 0 \end{pmatrix} \begin{pmatrix} 2 \\ 8 \\ 0 \end{pmatrix} \begin{pmatrix} 1 \\ 8 \\ 0 \end{pmatrix} \begin{pmatrix} 2 \\ 8 \\ 0 \end{pmatrix} \begin{pmatrix} 1 \\ j \\ 0 \end{pmatrix} \begin{pmatrix} 2 \\ j_8 \\ t_8 \end{pmatrix} \right), (0,0) \right)$$

with j_8 and t_8 chosen (non-uniquely) with $p(2, j_8, t_8) = j$. As a prelude to the main calculation, we prepare a list of $N+1$ initial states

$$\Psi = \{W^{t_k}\,\psi : \; k = 0, \ldots, N\}, \qquad t_0 = 0,$$

with precisely the time intervals needed for our moment calculations. The idea is to advance all of these states simultaneously, step by step,

$$\Psi \to W^s\,\Psi = \{W^{t_k+s}\,\psi : \; k = 0, \ldots, N\}, \qquad s = 0, \ldots, T_j(7) - 1,$$

computing the contributions to the right-hand side of (6.13) whenever $W^s\,\psi$ assumes the form $(V, (m,n))$ with

$$V = \begin{pmatrix} 0 \\ j_2 \\ t_2 \end{pmatrix}, \ldots, \qquad p(0, j_2, t_2) \in \{8, 10\}.$$

As discussed above, we expect there to be 468750 such events for $j = 1,7$ and 703125 of them for $j = 3,5$.

The results of our moment calculations for $1000 \leq t \leq 16000000$ are displayed in Figs. 6.10 and 6.11. The first figure shows a log-log plot of each of the moments versus the time interval t. The average slope for the kth moment is apparently consistent with power law $t^{k\nu}$ with exponent

$$\nu = \log 5 / \log 9$$

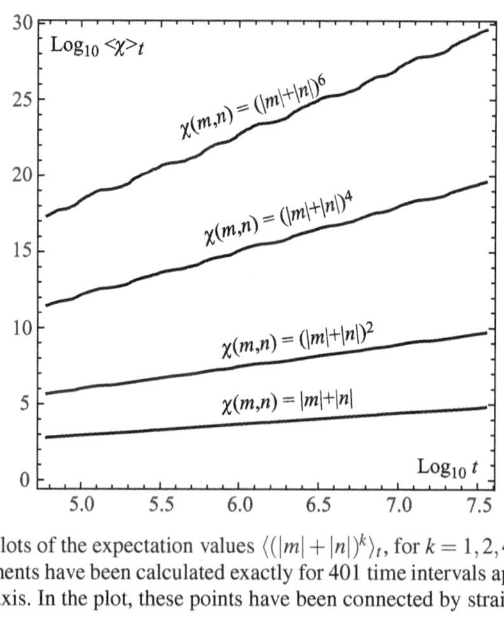

Fig. 6.10 Log-log plots of the expectation values $\langle(|m| + |n|)^k\rangle_t$, for $k = 1, 2, 4, 6$ and $36588 \leq t \leq 16000000$. The moments have been calculated exactly for 401 time intervals approximately equally spaced on the $\log t$ axis. In the plot, these points have been connected by straight line segments.

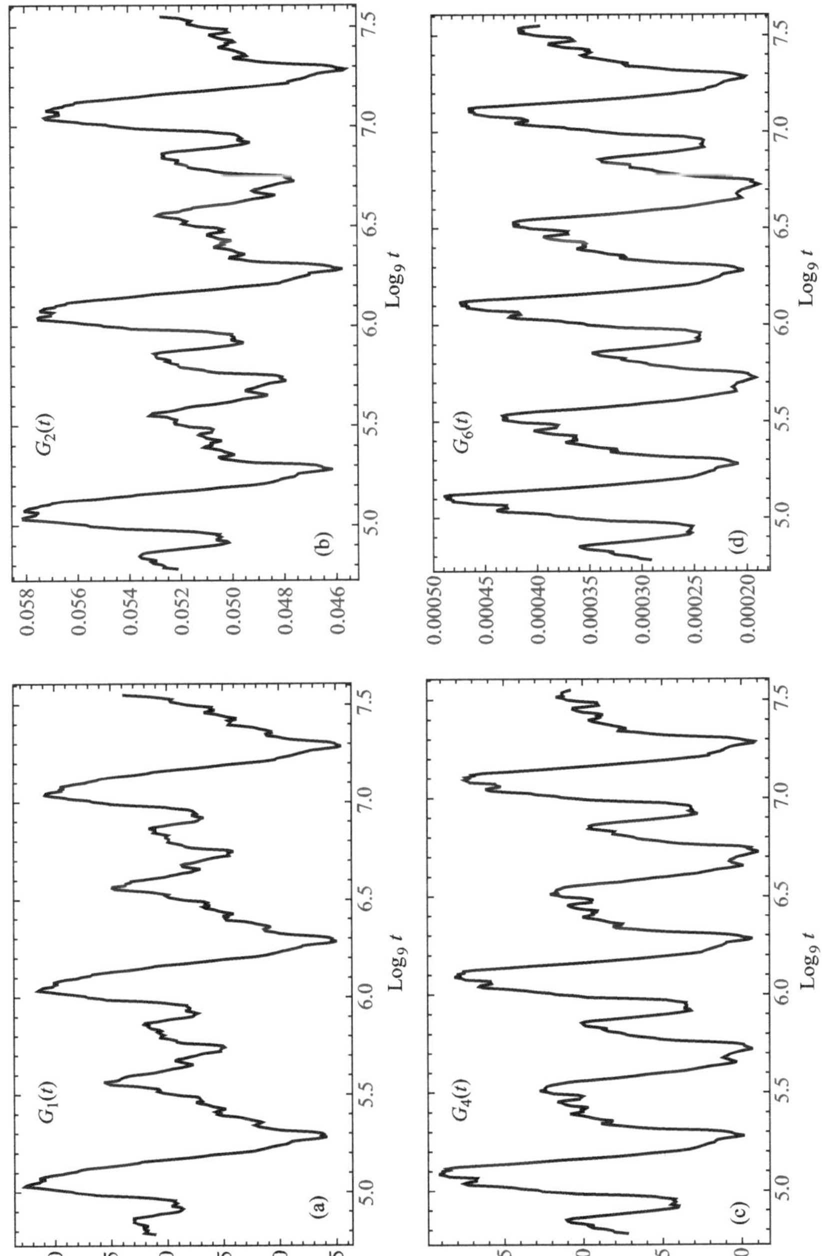

Fig. 6.11 Form factors $t^{-k\nu}\langle(|m|+|n|)^k\rangle_t$, for **(a)** $k = 1$, **(b)** $k = 2$, **(c)** $k = 4$, **(d)** $k = 6$, over the range $36588 \leq t \leq 16000000$.

found in Sect. 5.3, but there is clearly a log-periodic oscillation modulating the power law. In Fig. 6.11, we have divided out a factor t^{ν} to obtain a clear picture of the form factors with evident asymptotic unit periodicity with respect to $\log_9 t$.

A variant of the above computational experiment, which exploits the fact that the orbits are confined to a narrow strip containing the negative n axis, allows us to probe the evolving probability distribution in great detail. Using the same algorithm as before, with the same set of $N = 401$ time intervals, we have calculated

$$\langle \delta_n \rangle_t = \sum_m P((m,n),t).$$

plotting the resulting function of n for each of the selected t values. In each plot, we choose the scale of the n axis to be proportional to t^{ν} to facilitate the interpretation of the log-periodic structures found above for the kth moments. In Fig. 6.12, we show $\langle \delta_n \rangle_t$ for $t = 250000$. The probability distribution is characterized by a number

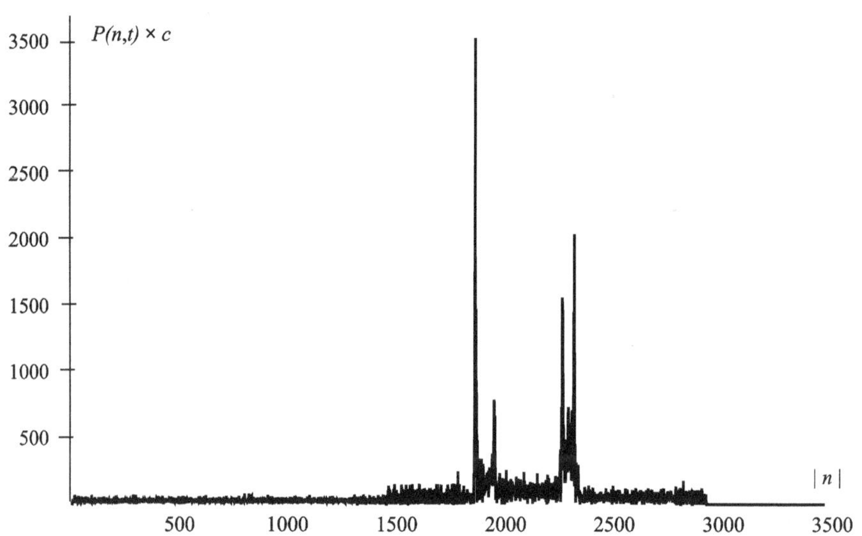

Fig. 6.12 Probability distribution $P(n,t)$ in the vertical lattice coordinate n, for $t = 250000$. The normalization factor is $c = 60 \times 5^5 = 187500$.

of narrow peaks which change their amplitudes and positions as t increases. Each prominent peak corresponds to significant trapping of the orbit within a small interval of the axis, sometimes for thousands of iterations. This effect is made evident in Fig. 6.13, where n has been plotted versus t for an orbit located in the region of peak probability at time $t = 250000$. A piece of this orbit leading up to trapping is shown, with successive points connected by line segments, in Fig. 6.14.

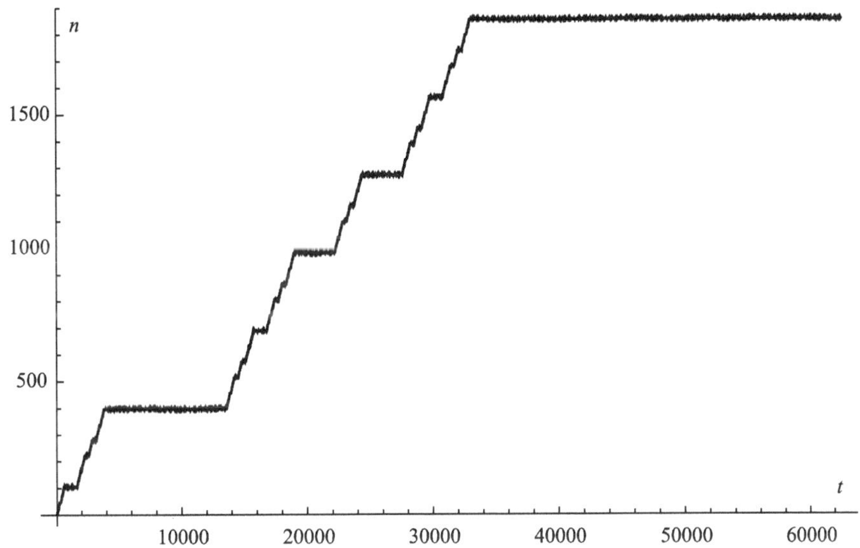

Fig. 6.13 Plot of n versus t for an orbit located in the highest peak of $P(n,t)$ in Fig. 6.12.

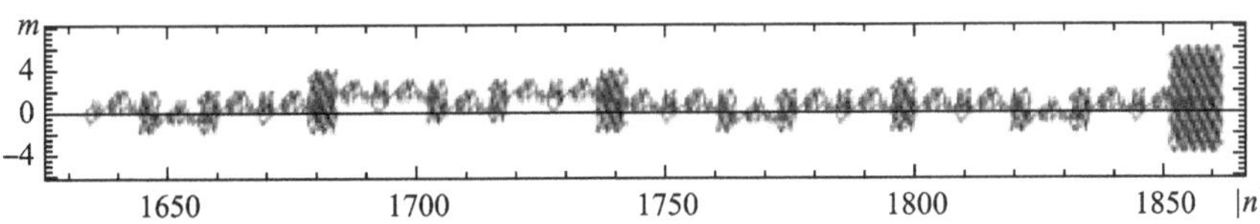

Fig. 6.14 Plot of the second half of the orbit of Fig. 6.13, rotated by 90°.

6.6 Discussion

Let us try to summarize what we have learned thus far about power-law asymptotic behavior in our class of kicked oscillator models. First of all, we note that power laws are associated with dynamically self-similar return map structures of the local map, namely *scaling sequences* of dressed domains. This means that in a specific model, there may multiple power laws, (at most) one associated with each independent scaling sequence. Whether a particular scaling sequence corresponds to a nontrivial power law is a rather delicate question, whose answer depends on the *synchronization*, or lack thereof, of the global 4-fold rotation number with the local return map periodicities. If there is such synchronization, we are able to construct a recursion matrix whose largest eigenvalue, ω_W, together with the temporal scale factor ω_T, determine the exponent of power law expansivity. The local geometric scale factor ω_K plays no role here.

In Chap. 5, we saw, both analytically and numerically, how global expansivity leads to an asymptotic power law for individual orbits, in the form of the existence of a lim-sup for the distance divided by t^{μ}, $\mu = \log \omega_W / \log \omega_T$. In Chap. 6, these results were supplemented by high-precision numerical calculations of the moments of a coarse-grained probability distribution. The numerical experiments provide strong evidence for the modulated power law

$$\langle r^k \rangle_t = t^{k\mu}(G_k(t) + O(t^{-\varepsilon}))$$

where r is either $|m| + |n|$ or $\sqrt{m^2 + n^2}$, $\varepsilon > 0$, and the form factor G_k is an eventually periodic function of $\log t$ with eventual period $\log \omega_T$. To prove this result analytically remains a challenging goal for future investigations.

References

Lowenstein J.H., Poggiaspalla G., and Vivaldi F. (2005) Sticky orbits in a kicked-oscillator model, *Dynamical Systems* **20**, 413–451.

Chapter 7
Hamiltonian Round-Off

In this final chapter, we shift our attention to a variation on the theme of a resonantly kicked one-dimensional harmonic oscillator, namely a model of an invertible, area-preserving round-off map (Vivaldi, 1994; Lowenstein et al., 1997; Lowenstein and Vivaldi, 1998, 2000; Lowenstein and Liu, 2003; Akiyama et al., 2008). Here the oscillator is represented stroboscopically by application of the generalized rotation

$$\Psi(\mathbf{x}) = \mathbf{C} \cdot \mathbf{x},$$

where

$$\mathbf{C} = \begin{pmatrix} \lambda & -1 \\ 1 & 0 \end{pmatrix},$$

and $\lambda = 2\cos(2\pi\rho)$ is a quadratic irrational and the rotation number ρ is assumed to be rational. The corresponding *round-off map* is defined, for integer x, y, to be

$$\Phi \begin{pmatrix} x \\ y \end{pmatrix} = \begin{pmatrix} \lfloor \lambda x \rfloor - y \\ x \end{pmatrix} \tag{7.1}$$

where $\lfloor \cdot \rfloor$ is the floor function. It is an easy exercise to verify that Φ is both invertible and the restriction to $\mathbb{Z}^2$ of an area-preserving map of the plane, and so it is natural to use the terminology *Hamiltonian round-off*. For a more detailed introduction to the subject, the reader is referred to the articles cited above.

7.1 Vector field

The generalized $2\pi p/q$ rotation Ψ returns to the initial point after q iterations, while the round-off map Φ ends up displaced by a vector

$$\mathbf{d}(\mathbf{x}) = \Phi^q(\mathbf{x}) - \mathbf{x},$$

the *round-off error*. A simple example with non-zero round-off error is illustrated in Fig. 7.1. In Fig. 7.2 we show the possible displacement vectors for the round-off models with $\rho = 1/5$ and $\rho = 5/12$, respectively.

One can interpret $\mathbf{d}(\mathbf{x})$ as a *vector field* on $\mathbb{Z}^2$, assigning a lattice displacement to each point x. Starting at a given $\mathbf{x}$ and iteratively applying Φ^q (resp. Φ^{-q}), one obtains the forward (resp. backward) orbit through $\mathbf{x}$. Thanks to the invertibility of Φ, the full orbits are disjoint. By drawing the vector field displacements for all the lattice points in a given region as line segments, one obtains a graphical representation of the orbits intersecting that region. Figs. 7.3 and 7.4 display such vector field diagrams for rotation numbers 1/5 and 5/12, respectively.

Obviously, our round-off model is closely related to the kicked oscillators of the preceding chapters with quadratic irrational parameters. One way of thinking about

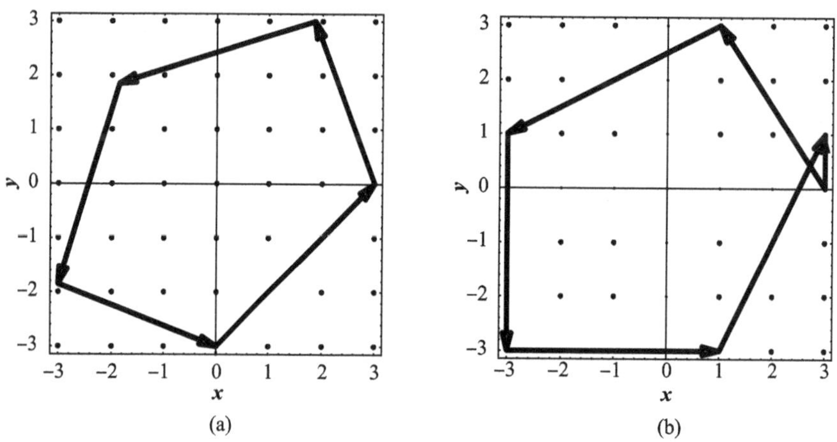

Fig. 7.1 Five iterations of (**a**) Ψ and (**b**) Φ for $p/q = 1/5$.

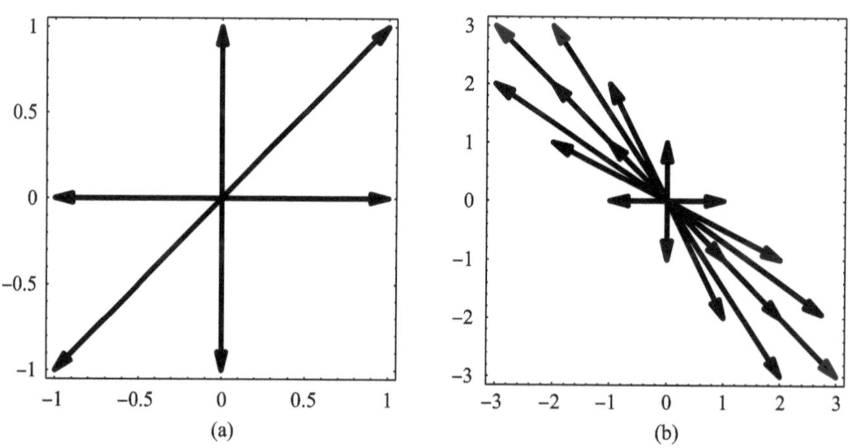

Fig. 7.2 Possible displacement vectors for (**a**) $p/q = 1/5$, (**b**) $p/q = 5/12$.

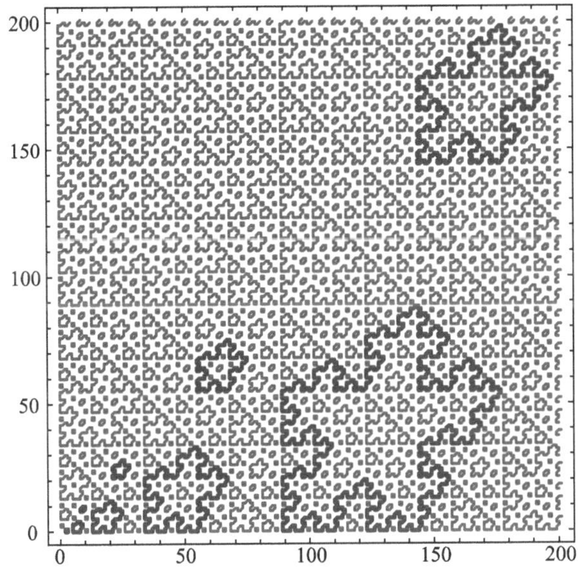

Fig. 7.3 Vector field diagram for $p/q = 1/5$. Only non-zero displacements have been drawn. Note the sequences of increasingly large "snowflakes" along the positive x axis and the main diagonal. Note also the large amount of white space, reflecting the prevalence of period-5 orbits with displacement vector $(0,0)$.

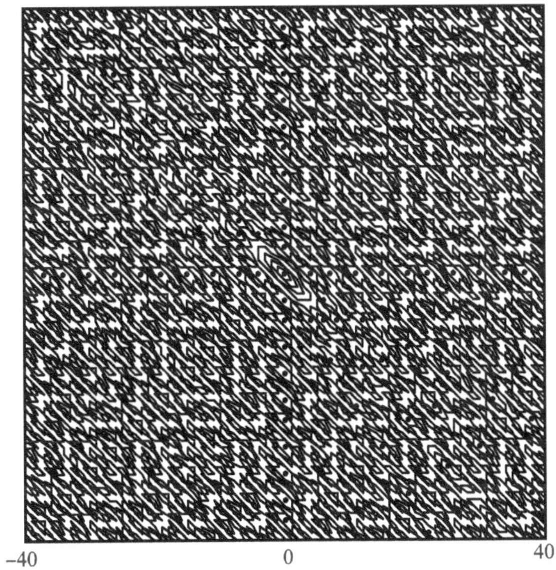

Fig. 7.4 Vector field diagram for $p/q = 5/12$. The period-12 orbits appear as dots which are too scarce to have a non-zero density.

the relation is the following. The round-off map is nothing but the lattice restriction of the map W of (1.3) with the true rotation matrix replaced by its oblique cousin $\mathbf{C}$ and the kick function $f(x)$ taken to be the fractional part of λx. Unfortunately there is a mismatch between the unit periodicity of $\mathbb{Z}^2$ and the periodicity $\tau = \lambda^{-1}$ of $f(x)$, and so the local-global decomposition introduced in Chap. 1 does not work. Nevertheless, as we will see below, the essence of the dynamics is once again to be found in a piecewise isometry of the square, with quadratic irrational parameter conjugate to λ. This *localization* of the round-off map on the integer lattice will allow us to take over without change the results of Chaps. 2 and 3 and quickly turn our attention to the issue of the propagation of round-off error on the infinite lattice.

7.2 Localization

The *localization map* $\mathscr{E}(\mathbf{x})$ maps the integer lattice $\mathbb{Z}^2$ into a dense subset of the square

$$\Omega = \begin{cases} [0,\tau)^2, & \tau > 0, \\ (\tau,0]^2, & \tau < 0. \end{cases}$$

Specifically,

$$\mathscr{E}((x,y)) = (x - m_0\tau, m_0 - y - m_1\tau), \qquad m_0 = \lfloor \lambda x \rfloor, \qquad m_1 = \lfloor \lambda(m_0 - y) \rfloor. \tag{7.2}$$

The following theorem shows the conjugacy of the round-off map Φ on $\mathbb{Z}^2$ to a piecewise isometry $\tilde{\Phi}$ on $\mathscr{E}(\mathbb{Z}^2) \subset \Omega$:

Theorem 7.1. *Let λ be a quadratic irrational of the form $\lambda = 2\cos 2\pi\rho$, ρ rational, with $\tilde{\lambda}$ its conjugate root. Let Φ be the round-off map with parameter λ. Further, let $\tilde{\Phi}$ be the piecewise isometry on Ω defined by*

$$\tilde{\Phi}(\zeta) = \tilde{\mathbf{C}} \cdot \zeta \bmod \Omega \stackrel{\text{def}}{=} \tilde{\mathbf{C}} \cdot \zeta - (0, m\tau), \tag{7.3}$$

where

$$\tilde{\mathbf{C}} = \begin{pmatrix} 0 & 1 \\ -1 & \tilde{\lambda} \end{pmatrix}$$

and m is the unique integer such that the right-hand side of (7.3) *lies in Ω. Then*

$$\Phi = \mathscr{E}^{-1} \circ \tilde{\Phi} \circ \mathscr{E}.$$

Proof. Thanks to the irrationality of τ, if ζ is of the form $x + m\tau$, $x, m \in \mathbb{Z}$, then the choice of x and m is unique. Hence $\mathscr{E}^{-1}$ is well defined on $\mathscr{E}(\mathbb{Z}^2)$ and it will be sufficient to verify the intertwining relation

$$\tilde{\Phi} \circ \mathscr{E}(\mathbf{x}) = \mathscr{E} \circ \Phi(\mathbf{x}) \tag{7.4}$$

for arbitrary $x, y \in \mathbb{Z}$. To evaluate the left-hand side, we note that

$$\tilde{\lambda} = \mu(q) - \lambda = (h(q) - 2)\tau, \qquad \mu(q), h(q) \in \mathbb{Z},$$

so that

$$\begin{aligned}
\tilde{\Phi} \circ \mathscr{E}(\mathbf{x}) &= \begin{pmatrix} 0 & 1 \\ -1 & \tilde{\lambda} \end{pmatrix} \begin{pmatrix} x - m_0\tau \\ m_0 - y - m_1\tau \end{pmatrix} \bmod \Omega \\
&= \begin{pmatrix} m_0 - y - m_1\tau \\ m_1 - x + \tau((h(q) - 2)(m_0 - y) - \mu(q)m_1 + m_0) \end{pmatrix} \bmod \Omega \\
&= \begin{pmatrix} m_0 - y - m_1\tau \\ m_1 - x \end{pmatrix} \bmod \Omega
\end{aligned}$$

But this is just $\mathscr{E}$ applied on (7.1), and so the lemma is proved. □

It is easy to invert the localization map, defining $\mathscr{E}^{-1}$ on the points of $\Omega \cap \mathbb{Z}[\tau]^2$:

$$\mathscr{E}^{-1}\begin{pmatrix} \zeta_0 \\ \zeta_1 \end{pmatrix} = \begin{pmatrix} [\zeta_0]_1 \\ -[\zeta_1]_1 - [\zeta_0]_\tau \end{pmatrix}, \tag{7.5}$$

where we define the "integer part" and "tau part" of $m + n\tau$, $m, n \in \mathbb{Z}$, by

$$[m + n\tau]_1 = m, \qquad [m + n\tau]_\tau = n.$$

We note that $\mathscr{E}^{-1}$ is additive and distributive with respect to multiplication by integers:

$$\mathscr{E}^{-1}(\boldsymbol{\chi} + \boldsymbol{\psi}) = \mathscr{E}^{-1}(\boldsymbol{\chi}) + \mathscr{E}^{-1}(\boldsymbol{\psi}), \qquad \mathscr{E}^{-1}(k\boldsymbol{\chi}) = k\mathscr{E}^{-1}(\boldsymbol{\chi}), \quad k \in \mathbb{Z}.$$

7.3 Localization of the vector field and periodic orbits

We now want to show how the localization map $\mathscr{E}$ provides a remarkable organization of the lattice orbits according to shape (a concept yet to be defined precisely). We begin by calculating the possible vector field vectors $\mathbf{d}^{(i)}$, $i = 0, 1, \ldots, N$ for a given model, showing that the set of lattice points assigned a specific vector field $\mathbf{d}^{(i)}$ is mapped by $\mathscr{E}$ into a finite number of convex polygons in Ω. The simplest way of establishing these results is to construct the time-q dynamical partition of Ω for the piecewise isometry $\tilde{\Phi}$ (with parameter $\tilde{\lambda}$) defined in Theorem 7.1. Recall from Sect. 6.1 that this is a partition of Ω into convex polygons $D^{(k)}$, each of which is mapped without branching for q iterations of the local map. The cumulative effect of these mappings on $D^{(k)}$ is an isometry, namely

$$\tilde{\Phi}(\boldsymbol{\zeta}) = \tilde{\mathbf{C}}^q \cdot \boldsymbol{\zeta} + \boldsymbol{\delta}^{(k)} = \boldsymbol{\zeta} + \boldsymbol{\delta}^{(k)}, \quad \boldsymbol{\delta}^{(k)} \in \mathbb{Z}[\tau],$$

where we have made use of the identity $\tilde{\mathbf{C}}^q = \mathbf{1}$. Now the vector field $\mathbf{d}$ assigned to any lattice point $\mathscr{E}^{-1}(\boldsymbol{\zeta})$, $\boldsymbol{\zeta} \in D^{(k)}$, is by definition given by

$$\mathbf{d}(\mathscr{E}^{-1}(\boldsymbol{\zeta})) = \mathscr{E}^{-1}(\boldsymbol{\zeta} + \boldsymbol{\delta}^{(k)}) - \mathscr{E}^{-1}(\boldsymbol{\zeta}).$$

Using the additivity of $\mathscr{E}^{-1}$, we get, finally,

$$\mathbf{d}(\mathscr{E}^{-1}(\boldsymbol{\zeta})) = \mathscr{E}^{-1}(\boldsymbol{\delta}^{(k)}) = \begin{pmatrix} [\delta_0^{(k)}]_1 \\ -[\delta_1^{(k)}]_1 - [\delta_0^{(k)}]_\tau \end{pmatrix}.$$

This establishes the following theorem:

Theorem 7.2. *Suppose the convex polygon D has the property that it is mapped without branching during q iterations of* $\tilde{\Phi}$. *Then the vector field assumes a unique value on the set* $\mathscr{E}^{-1}(D \cap \mathbb{Z}[\tau]^2)$.

The localization of the vector field is illustrated in Fig. 7.5, for $p/q = 1/5$. Above, one sees a 41×41 array of lattice points, color-coded according to the vector field values. Below, the local images of these points in Ω are displayed, showing the localization of the vector field (arrows) within the convex polygons of the time-q dynamical partition.

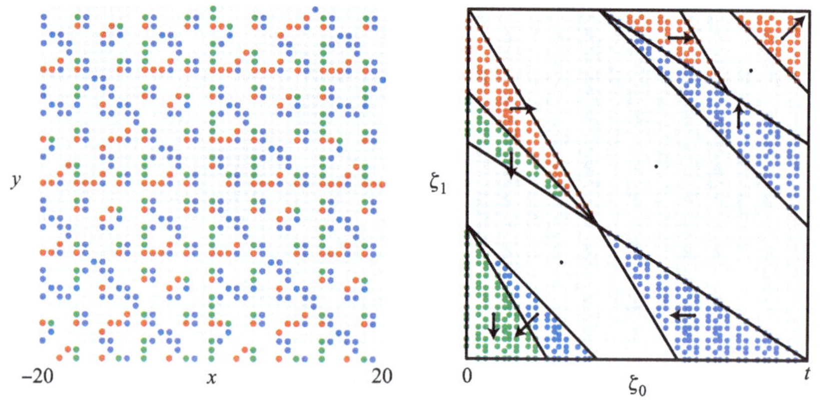

Fig. 7.5 Localization of the vector field for $p/q = 1/5$. (See the color figure at the end of the book.)

From the localization of the vector field, it is a relatively simple step to the localization of *walks* , i.e. dynamically generated vector-field sequences

$$\{\mathbf{d}^{(k_0)}, \mathbf{d}^{(k_1)}, \ldots, \mathbf{d}^{(k_{N-1})}\}.$$

Here we extend the notion of a walk to include all iterations of Φ, which means that, except very close to the origin, a walk will include q disjoint branches, each of which is a walk in the conventional sense, i.e. a sequence of small steps forming a broken line segment on the lattice. Two walks will be considered to *have the same shape* if their vector-field sequences are identical, which is easily seen to be equivalent to congruence of their respective branches as broken line segments.

Consider now the $\mathscr{E}$ images of all the lattice points (including the initial point, but not the final one) of an N-step walk. These clearly will inhabit convex subpolygons of the local vector-field domains $D^{(k)}$. For a three-step walk,

$$\{\mathbf{d}^{(k_0)}, \mathbf{d}^{(k_1)}, \mathbf{d}^{(k_2)}\},$$

the polygons would be

$$P_1 = \Phi^{-2}(\Phi(\Phi(D^{(k_0)}) \cap D^{(k_1)}) \cap D^{(k_2)})\}, \qquad P_2 = \Phi(P_1), \qquad P_3 = \Phi(P_2).$$

Probably the most significant walks to be considered are the periodic orbits, since in the models treated in this chapter the points on periodic orbits constitute the full set of points on the lattice (for $p/q = 1/5$, this is strongly suggested by Fig. 7.3). The localization of periodic orbits in the $p/q = 1/5$ model is nicely illustrated in Fig. 7.6. On the left, we see all points of the period-20 orbits (5 disjoint branches, each with

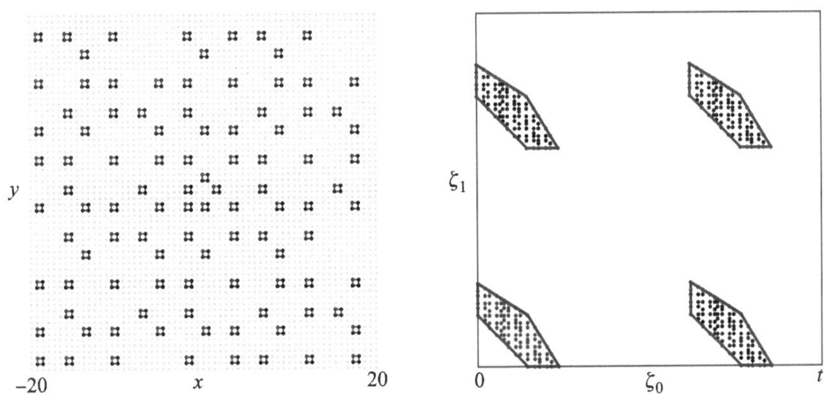

Fig. 7.6 Localization of the period-20 orbits for $p/q = 1/5$.

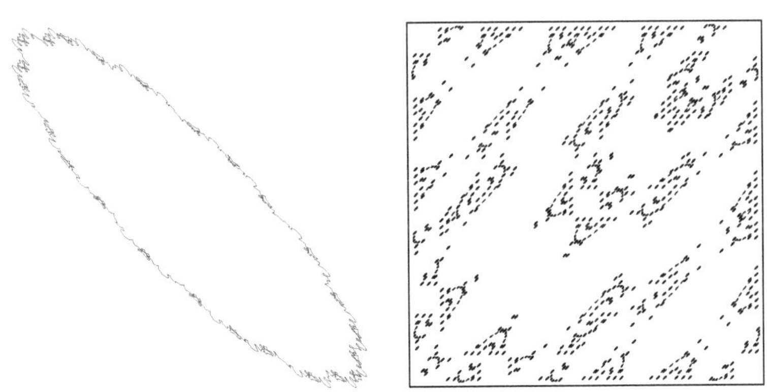

Fig. 7.7 Localization of the period-9444 orbits for $p/q = 5/12$.

4 points arranged in a square) in a 60×60 portion of $\mathbb{Z}^2$ centered at the point $(0,0)$. Each walk is now a sequence of four different displacement vectors repeated ad infinitum, and there are infinitely many walks with the same shape distributed fairly uniformly over the plane. On the right is the set of $\mathscr{E}$-images of the lattice points on the left, neatly segregated into four disjoint "regular" pentagons, each one associated with one of the vector field types. The q-fold symmetry of the polygons, visited q times per period, is typical of this model, but is not a general rule (in another case, a period-20 orbit might have a local orbit of 20 unsymmetrical polygons, each visited once per period). A much more complicated orbit, in both its global and local versions, is shown in Fig. 7.7.

7.4 Symbolic codes for walks

In our treatment of transport in kicked oscillator models, we found it useful to think of the local piecewise isometries and related global lattice permutations as two manifestations of the same symbolic dynamics. Given the existence of the localization map $\mathscr{E}$, it is rather obvious that the lattice dynamics of the round-off model with the same local map $\tilde{\boldsymbol{\Phi}}$ must somehow be a third representation of the same symbolic scheme. Let us now examine how this works in detail.

The simplest coding of local-map orbits is provided by the generating partition of the square Ω into subdomains Ω_i. Keeping track of the sequence $\{i_0, i_1, \ldots\}$of the Ω_{i_k} through which an orbit passes is an effective way of describing its local motion. For the kicked-oscillator models, this sequence uniquely specified a lattice itinerary as well. Unfortunately, the same is not true of the round-off model's lattice motions: for $\boldsymbol{\zeta}_1$ and $\boldsymbol{\zeta}_2$ in the same Ω_i, the vector-field displacements associated with the corresponding lattice points $\mathscr{E}^{-1}(\boldsymbol{\zeta}_1)$ and $\mathscr{E}^{-1}(\boldsymbol{\zeta}_2)$ are not necessarily the same.

A more useful coding corresponds to the finer partition provided by the q-step dynamical partition of Ω, i.e. the partition into maximal convex polygons Ξ_i which are mapped forward q steps without intersecting a discontinuity line. By construction, the vector field is uniform on each domain of the partition, although there may be more than one such domain with the same vector field. Now, to each sequence of Ξ_i, the shape of all lattice walks whose local representatives pass through those domains is uniquely determined. Here the concept of shape is in the generalized sense introduced in Sect. 7.3.

The recursive tiling property of the local map in our renormalizable kicked-oscillator models allowed us to introduce a highly efficient hierarchical coding and symbolic dynamics. Much of this can be taken over to the formulation of round-off models, with an important difference. In the models we study, all lattice points are localized within periodic polygons, and so the intersection of $\mathscr{E}(\mathbb{Z}^2)$ with the residual set of the local map is empty. Nevertheless, for mappings of tiles within the hierarchy, the symbolic dynamics remains a valuable tool.

7.5 Construction of the probability distribution

Theorem 7.1 establishes the dynamical equivalence of round-off maps and piecewise isometries of the square for quadratic irrational parameters. It allows us to take over immediately everything we have learned about renormalizability and Vershik-type symbolic dynamics. In some ways the situation is analogous to that of the kicked-oscillator models which have been our major concern so far. But there are crucial differences, especially when it comes to formulating a meaningful transport problem. In both cases, we are interested in a probability distribution $P(\mathbf{x},t)$ giving the probability that the orbit of a randomly selected initial point with lattice coordinates $\mathbf{x}_0$ ends up displaced on the lattice by $\mathbf{x}$ after t iterations of the elementary map. To remove the effects of global rotations, one adopts a stroboscopic viewpoint, requiring that t be a multiple of 4 (resp. q) in the kicked-oscillator (resp. round-off) case.

In the kicked-oscillator case, the local phase space is the fundamental domain of an infinite crystalline array, and so, using translation invariance, random selection of the initial point refers naturally to the uniform distribution of points in the square, and convex polygons are assigned weights in the probability calculation proportional to their respective areas. In the round-off case, the phase space consists solely of the lattice points, and the local square Ω is an abstraction. Moreover, we do not have translation invariance: the trajectories initiated at different lattice sites can have different shapes. Nevertheless, it is possible to assign nonzero weights to lattice orbits according to their shapes, and moreover to identify these weights with the areas of convex polygons in Ω. This will then allow us to construct $P(\mathbf{x},t)$ along the same lines as we used in Chapter 6 for the kicked oscillator models.

To understand how a probability measure can be assigned to a lattice, let us consider the analogous one-dimensional problem (Lowenstein and Vivaldi, 1998). The points of the non-negative integer lattice $\mathbb{Z}_+ = \{0,1,\ldots\}$ can be "localized" on the unit interval by the map

$$\eta : \mathbb{Z} \to [0,1), \qquad \eta(m) = \{\lambda m\} = \lambda m - \lfloor \lambda m \rfloor,$$

where λ is a positive irrational real number and $\{\cdot\}$ is the fractional part function. For any interval $I \subset [0,1)$, the set $\eta^{-1}(I) \subset \mathbb{Z}_+$ can be assigned a density

$$\mathscr{D}(\eta^{-1}(I)) = \lim_{n\to\infty} \frac{1}{n} \#\{m \in \eta^{-1}(I) : m < n\} \leq 1.$$

By Weyl's theorem (see Hardy and Wright, 1979, p. 445), this density is equal to the length of I,

$$\mathscr{D}(\eta^{-1}(I)) = |I|.$$

It is a standard exercise in measure theory to promote the density function on special subsets to a measure on the lattice $\mathbb{Z}_+$. The latter is additive over finite unions of disjoint sets, but not over countable unions. For example, the set $\mathbb{Z}_+$ has unit measure, but is the union of countably many discrete integers, each of which has zero measure.

The reader can verify that in our round-off models, the localization map $\mathscr{E}$ is essentially the 2-dimensional analogue of our 1-dimensional example. It is then straightforward to generalize the above arguments and prove (Lowenstein and Vivaldi, 1998) the following lemma.

Lemma 7.1. *Let $D \subset \Omega$ be a convex polygon. Then the density*

$$\mathscr{D}(\mathscr{E}^{-1}(D)) = \lim_{n\to\infty} \frac{1}{4n^2} \#\{(l,m) \in \mathscr{E}^{-1}(D) : -n < l,m < n\} \leq 1.$$

is equal to the area of D divided by the area of Ω.

Note that this result extends immediately to finite unions of convex polygons.

Having defined a density measure on the lattice, we are in a position immediately to define the probability distribution for the propagation of round-off error, namely

$$P(\mathbf{x},t) = \mathscr{D}(\{\mathbf{z} \in \mathbb{Z}^2 : \Phi^t(\mathbf{z}) - \mathbf{z} = \mathbf{x}\}), \tag{7.6}$$

with t restricted to integer multiples of q. Since the argument of $\mathscr{D}$ is the intersection of a finite set of walks, namely all those which produce a net lattice displacement of $\mathbf{z}$ after t iterations, the localized image of that set resides in a finite set of convex polygons in Ω. By Lemma 7.1, the right-hand side of (7.6) is just the total area of those polygons.

For an efficient calculation of the probability distribution and its moments, we can turn to the procedures of Sect. 6.1 which exploit both recursive tiling and hierarchical symbolic dynamics. There are, however, some differences worth noting. In the kicked-oscillator case, we had two alternative ways of implementing the local-global correspondence. In the simpler version, which we adopted, each application of the map W corresponds both to the a unique piecewise isometry of the square and to a lattice isometry with rotation number $1/4$, whose displacement vector depends uniquely on the domain Ω_i containing the local point. Iteration of W produces a quadruply branched generalized walk on the lattice. An alternative implementation focuses attention on the fourth-iterate map W^4, whose generating domain map involves a finer partition of Ω. The local representation is again a piecewise isometry of Ω, but now the corresponding lattice isometry is a translation, again depending on the local subdomain. Now iteration of W produces four singly branched walks, one for each value of the rotational phase.

In the round-off models, we have no local-global decomposition of the map Φ, only of Φ^q, and so only the second scenario is available to us. Thus in our probability calculations we are obliged to use a lowest-level partition in which the vector field vector is unique on each subdomain, and to keep track of q separate walks simultaneously if we want to include a complete set of initial conditions. Fortunately, these add only minor complications to our algorithms.

7.6 Rotation number 1/5

In the remainder of this chapter we will consider several interesting examples, starting with $p/q = 1/5$ and $\lambda = (\sqrt{5}-1)/2$, the one case in the literature where rigorous results concerning the long-time behavior of the moments have been obtained (Lowenstein and Vivaldi, 1998). Here we follow closely the treatment of that article.

Let us review briefly what we have learned about this model so far. The round-off map with $\lambda = (\sqrt{5}-1)/2$ is defined in (7.1). Its vector field, $\mathbf{d}(\mathbf{x}) = \Phi^5(\mathbf{x}) - \mathbf{x}$, is plotted, over a 200×200 portion of the lattice, in Fig. 7.3. We note the apparent absence of unbounded orbits (proved for the entire lattice by Lowenstein et al. (1997) and generalized to other models by Akiyama et al. (2008), as well as the presence of scaling sequences of periodic orbits (apparently tending toward fractal snowflakes) arrayed along the x axis and the $45°$ diagonal. The localization of the vector field is illustrated in Fig. 7.5.

7.6.1 Recursive tiling for the local map

According to Theorem 7.1, the local map is given by

$$\tilde{\Phi}(\zeta) = \tilde{\mathbf{C}} \cdot \boldsymbol{\zeta} \mod \Omega,$$

with

$$\Omega = [0,\tau)^2, \qquad \tilde{\mathbf{C}} = \begin{pmatrix} 0 & 1 \\ -1 & -\tau \end{pmatrix}, \qquad \tau = \lambda^{-1} = \frac{\sqrt{5}+1}{2}.$$

This is the same as the local map of Lowenstein and Vivaldi (1998), apart from a trivial interchange of the coordinates ζ_0 and ζ_1.

The generating domain and its four atoms, shown in Fig. 7.8, are

$$\begin{aligned}
D(\Omega) &= \mathscr{D}((0,1,0,1),(-1,-1,1,1),(0,0,\tau,\tau)),\\
D_0(\Omega) &= \mathscr{D}((0,1,2),(-1,-1,-1),(0,0,0)),\\
D_1(\Omega) &= \mathscr{D}((0,2,1,2),(-1,1,-1,-1),(0,0,0,-\tau)),\\
D_2(\Omega) &= \mathscr{D}((0,1,2,0,2),(-1,1,-1,1,1),(0,\tau,-2\tau,\tau,-\tau)),\\
D_3(\Omega) &= \mathscr{D}((0,1,2),(1,1,1),(\tau,\tau,-2\tau)).
\end{aligned}$$

The choice of level-0 scaling domain is not unique, but it is advantageous to follow Lowenstein and Vivaldi (1998) (rather than Kouptsov et al., 2002) and define

$$D(0) = \mathscr{D}((0,2,3),(1,-1,-1),(0,0,1-\tau)).$$

The level-zero subdomains $D_j(0)$ and their return times $T_j(0)$ are

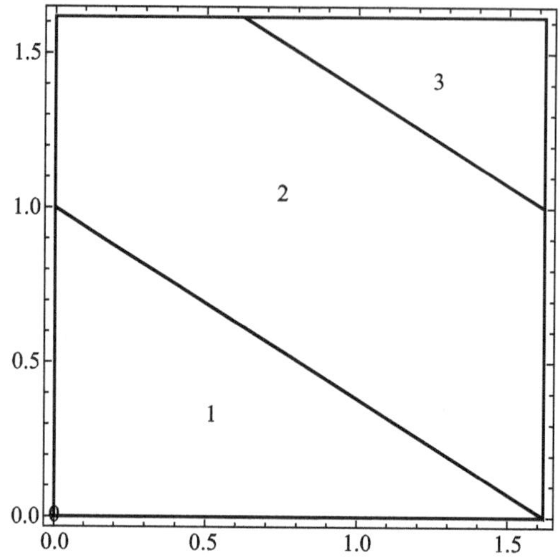

Fig. 7.8 Generating partition for $p/q = 1/5$.

$$D_0(0) = \mathscr{D}((1,0,2,1,3),(-1,1,-1,1,-1),(0,0,0,-1+\tau,1-\tau)), \quad T_0(0) = 4$$
$$D_1(0) = \mathscr{D}((0,1,3),(1,1,-1),(0,0,1-\tau)), \quad T_1(0) = 6.$$

Here the coordinates are relative to the scaling limit point $(\tau,1)$.

The level-0 tiling, including not only the return orbits of $D_0(0)$ and $D_1(0)$, but also the three period-1 pentagons

$$\Pi_0(\Omega) = \mathscr{D}((0,2,4,1,3),(-1,-1,-1,-1,-1),(0,-\tau,-\tau,0,1)),$$
$$\Pi_1(\Omega) = \mathscr{D}((1,4,2,0,3),(1,1,1,1,1),(\tau,-\tau,-\tau,\tau,2+\tau)),$$
$$\Pi_2(\Omega) = \mathscr{D}((0,2,4,1,3),(1,1,1,1,1),(\tau,-2\tau,-2\tau,\tau,3+\tau)),$$

is shown in Fig. 7.9, together with the level-1 tiling of $D(0)$. Note that the limit point of the scaling sequence is $(\tau,1)$, and that the level-1 scaling triangle $D(1)$ straddles the boundary between $D_0(0)$ and $D_1(0)$. This is different from our usual convention that $D(L)$ lies intirely in a single $D_j(L-1)$. We note that the level-1 tiling includes two periodic pentagons, $\Pi_0(0)$ and $\Pi_1(0)$, where, expressed in terms of coordinates relative to the scaling limit,

$$\Pi_0(0) = \mathscr{D}((1,4,2,0,3),(-1,-1,-1,-1,-1),(0,2-\tau,0,1-\tau,1-\tau)),$$
$$\Pi_1(0) = \mathscr{D}((0,3,1,4,2),(1,1,1,1,1),(0,3-2\tau,0,2-\tau,2-\tau)),$$

with minimal periods 14 and 26, respectively. The higher-level members of each scaling sequence are obtained by rescaling of the coordinates relative to $(\tau,1)$ by powers of $\tau^{-2} = 2-\tau$.

Once again referring to Fig. 7.9, we see that the return orbit of $D_0(1)$ visits $D_0(0)$ twice and $D_1(0)$ once, while that of $D_1(1)$ visits those domains 2 and 3 times, re-

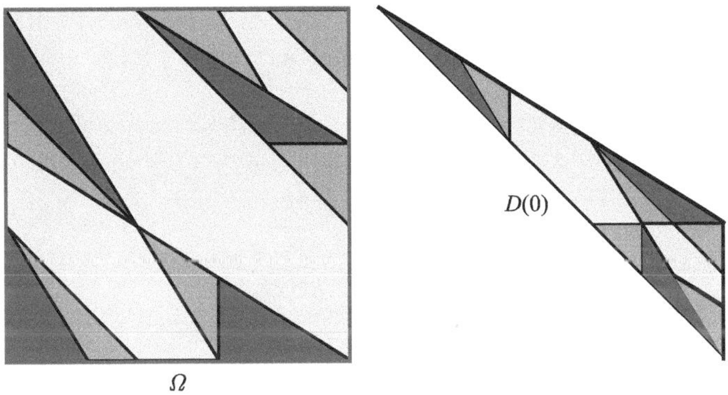

Fig. 7.9 Left: level-0 tiling of Ω. Right: level-1 tiling of $D(0)$.

spectively. Thus the incidence matrix is

$$\mathbf{A} = \begin{pmatrix} 2 & 2 \\ 1 & 3 \end{pmatrix},$$

and we get for the level-L return times,

$$(T_0(L), T_1(L)) = (T_0(0), T_1(0))\mathbf{A}^L = \left(\frac{1}{3}(10 \times 4^L + 2), \frac{2}{3}(10 \times 4^n - 1)\right). \quad (7.7)$$

The usefulness of recursive tiling for describing lattice walks is already evident at level 0. Comparison of Fig. 7.9 with Fig. 7.5 shows that every domain in the level-0 tiling is assigned a unique vector-field displacement. This is a property which is not shared by the original generating partition, but is now guaranteed for all higher-level tilings.

7.6.2 Probability distribution $P(\mathbf{x}, t)$

The probability distribution $P(\mathbf{x}, t)$ is normalized to unity for all t, and is initially entirely concentrated at the origin,

$$P(\mathbf{x}, 0) = \begin{cases} 1 & \mathbf{x} = (0,0), \\ 0 & \mathbf{x} \neq (0,0). \end{cases}$$

The relative frequencies on the lattice of the seven vector field displacements are given by $P(\mathbf{x}, 5)$ (recall the constraint that t be a multiple of 5):

$$P(\mathbf{x}, 5) = \begin{cases} 14\tau - 22 = 0.652\ldots & \mathbf{x} = (0,0), \\ (5 - 3\tau)/2 = 0.073\ldots & \mathbf{x} = \pm(1,0), \pm(0,1), \\ (13 - 8\tau)/2 = 0.028\ldots & \mathbf{x} = \pm(1,1). \end{cases}$$

As is evident in Fig. 7.3, nearly two thirds of the lattice points are fixed points of Φ^5. They form an inert background which does not participate in the time evolution of $P(\mathbf{x},t)$.

As we saw in Chap. 6, a simple way of calculating $P(\mathbf{x},t)$ for small t is provided by straightforward calculation of the dynamical partition of Ω for a given t. Each convex polygonal domain of the partition corresponds to a distinct displacement vector $\mathbf{x}_i$ and a distinct area A_i. One then has

$$P(\mathbf{x},t) = \tau^{-2} \sum_i A_i \delta(\mathbf{x},\mathbf{x}_i),$$

where $\delta(\mathbf{a},\mathbf{b})$ is equal to 1 for equal arguments and vanishes otherwise. Unfortunately, this technique quickly becomes unwieldy for increasing t, and so once more we turn to recursive tiling for relief. From (6.7) and (6.8), we have

$$P(\mathbf{x},t) = P_{\text{fix}} + \tau^{-2} \sum_{L=0}^{L_*(t)-1} \sum_{j=0}^{1} \mathscr{A}(\Pi_j(L)) Q(\Pi_j(L),\mathbf{x},t) \qquad (7.8)$$
$$+\tau^{-2} \sum_{j=0}^{1} \mathscr{A}(D_j(L_*(t))) Q(D_j(L_*(t)),\mathbf{x},t)$$

where $Q(D,\mathbf{x},t)$ counts the number of times that a forward t-step displacement vector $\mathbf{x}$ is encountered along the return orbit of a tile D. Moreover, if T is the length of the return orbit, then $L_*(t)$ is chosen to be sufficiently high that each set $\tilde{\Phi}^k D$, $k = 0,1,\ldots,T-1$ has a unique forward t-step displacement vector. Explicitly we can choose (see (6.4) and (6.5))

$$L_*(t) \geq \min\{L : t \leq T_*(L)\}, \qquad T_*(L) = \sum_{L'=0}^{L-1} T_0(L').$$

Here we have been careful to choose, on the right-hand side of the last equation, the smaller of the two return times T_0 and T_1. Inserting (7.7), we have

$$T_*(L) = \frac{1}{3} \sum_{L'=0}^{L-1} (10 \cdot 4^{L'} + 2).$$

Summing the geometric series, we find

$$T_*(L) = \frac{10}{9}(4^L - 1) + \frac{2}{3}L, \qquad (7.9)$$

and so, for $L > 1$, we may choose

$$L_*(t) = \lceil \log_4(9t/10) \rceil. \qquad (7.10)$$

A simplifying feature of the chosen recursive tiling is apparent in Fig. 7.9: for $t \leq T_j(L) + T_*(L)$ iterations, the itineraries of $D_j(L)$ and $\Pi_j(L)$ are identical, and hence

$$Q(D_j(L_*(t)),\mathbf{x},t) = Q(\Pi_j(L_*(t)),\mathbf{x},t).$$

This allows us to write (7.8) in a more concise form,

$$P(\mathbf{x},t) = P_{\text{fix}} + \sum_{n=0}^{2L_*(t)} b_n P_n(\mathbf{x},t) \tag{7.11}$$

where

$$P_{2L-1}(\mathbf{x},t) = Q(\Pi_0(L),\mathbf{x},t)/T_0(L), \quad P_{2L}(\mathbf{x},t) = Q(\Pi_1(L),\mathbf{x},t)/T_1(L),$$

and

$$b_{2L-1} = \tau^{-2}\mathscr{A}(\Pi_0(L))T_0(L), \quad b_{2L} = \tau^{-2}\mathscr{A}(\Pi_1(L))T_1(L),$$
$$b_{2L_*(t)-1} = \tau^{-2}\mathscr{A}(D_0(L_*(t)))T_0(L_*(t)), \quad b_{2L_*(t)} = \tau^{-2}\mathscr{A}(D_1(L_*(t)))T_1(L_*(t)).$$

7.6.3 Fractal snowflakes

Figure 7.3 strongly suggests that the scaling sequences of periodic orbits in the $p/q = 1/5$ round-off model approach, when suitably rescaled, some sort of fractal curve as the periods tend to infinity. What is less apparent is that this geometric self-similarity will reflect itself in asymptotic scaling properties of the probability distribution $P(\mathbf{x},t)$ and its moments. Following Lowenstein and Vivaldi (1998), we will now show that both geometric and dynamical self-similarity are in fact valid: there is a precise relation between asymptotically long periodic orbits and fractal snowflakes obtained by a Koch construction, and this geometrical property underlies the modulated power-law behavior of moments.

We begin by showing that the periodic orbits can themselves be assembled, without approximation, by a Koch-type construction. The key concept here is that of a *walk* W, represented abstractly as a coded sequence of displacement vectors, i.e.

$$W = (\mathbf{d}_{c_0}, \mathbf{d}_{c_1}, \ldots, \mathbf{d}_{c_{N-1}})$$

where the vectors $\mathbf{d}_i$ are taken from a *displacement list* D of M vectors, and the c_i are integers in $0,\ldots,M-1$. If the set D is considered fixed, it is convenient to represent W by the *code sequence*

$$C = (c_0, c_1, \ldots, c_{N-1}).$$

The kth *vertex* $\mathbf{r}_k$ of a walk W is defined as the position (relative to the starting point) reached after k steps:

$$\mathbf{r}_0(W) = (0,0), \qquad \mathbf{r}_k(W) = \sum_{j=0}^{k-1} \mathbf{d}_{c_j}.$$

We denote by $\mathscr{W}(D)$ the set of all walks with displacement list D. A *substitution* ϕ maps each element of $\{0,\ldots,M-1\}$ into a finite sequence of elements of $\{0,\ldots,M'-1\}$, and extends naturally, by concatenation, to a map on code sequences.

To illustrate the usefulness of walks in the construction of fractal curves, let us consider the well known Koch snowflake. Here the displacement list D consists of six elementary steps,

$$\mathbf{d}_k = \left(\cos\frac{2\pi k}{6}, \sin\frac{2\pi k}{6}\right), \quad k = 0,\ldots,5.$$

We start with the three-step walk around an equilateral triangle,

$$W_0 = (\mathbf{d}_0, \mathbf{d}_2, \mathbf{d}_4).$$

Applying the code substitution

$$\phi(k) = (k, (k+5)_6, (k+1)_6, k), \quad k = 0,\ldots,5,$$

to the sequence $(0,2,4)$ produces a new sequence

$$\phi(0,2,4) = (0,5,1,0) \smile (2,1,3,2) \smile (4,3,5,4) = (0,5,1,0,2,1,3,2,4,3,5,4),$$

and a new walk (around a 12-sided hexagram)

$$W_1 = (\mathbf{d}_0, \mathbf{d}_5, \mathbf{d}_1, \mathbf{d}_0, \mathbf{d}_2, \mathbf{d}_1, \mathbf{d}_3, \mathbf{d}_2, \mathbf{d}_4, \mathbf{d}_3, \mathbf{d}_5, \mathbf{d}_4).$$

Iterating the process, shown geometrically in Fig. 7.10, we obtain at the nth repeti-

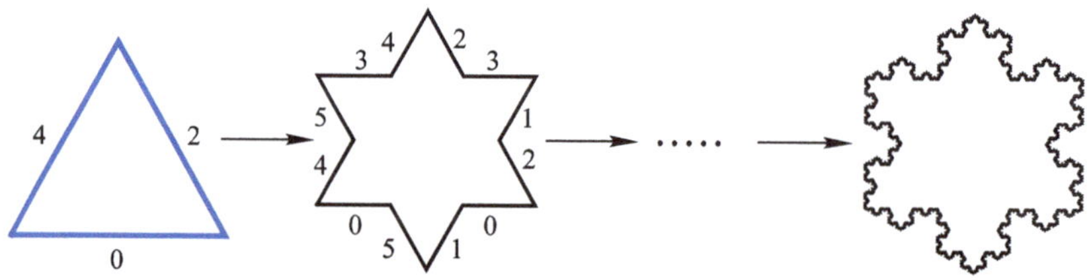

Fig. 7.10 Iterative construction of the Koch snowflake.

tion an $N_n = 3 \cdot 4^n$-sided polygon with diameter $d_n = 2 \cdot 3^{n-1/2}$, traversed counter-clockwise. Since

$$\frac{N_n}{N_0} = \left(\frac{d_n}{d_0}\right)^{\delta}, \qquad \delta = \frac{\log 4}{\log 3},$$

we see that in the limit $n \to \infty$, the polygons, rescaled by 3^{-n}, approach a hexagonally symmetric fractal of dimension δ.

A pentagonal Koch-type walk K_n resembling our round-off walks may be constructed using the displacement lists

$$D_- = ((0,-\lambda^2),(\lambda^2,0),(\lambda^3,\lambda^2),(-\lambda^3,\lambda^3),(-\lambda^2,-\lambda^3)), \tag{7.12}$$
$$D_+ = -\lambda D_+ = ((0,\lambda),(-\lambda,0),(-\lambda^2,-\lambda),(\lambda^2,-\lambda^2),(\lambda,\lambda^2)). \tag{7.13}$$

and the substitutions $\phi_\pm$ defined by

$$\left.\begin{array}{l}\phi_-(k) = (k,(k+1)_5,(k+4)_5,k)\\ \phi_+(k) = (k,(k+4)_5,(k+1)_5,k)\end{array}\right\}, \quad k = 0,\dots,4.$$

Starting with the walks on pentagons

$$K_{0\pm} = (d_0(D_\pm),d_1(D_\pm),d_2(D_\pm),d_3(D_\pm),d_4(D_\pm)),$$

the higher-level walks are generated by

$$K_{2l-1} = \sigma_-^l(K_{0-}), \qquad K_{2l} = \sigma_+^l(K_{0+}), \qquad l = 1,2,\dots.$$

where $\sigma_\pm$ are the maps induced by the substitutions $\phi_\pm$.

With suitable rescaling, the polygons traced out by the two sequences of walks can be made to have a single fractal limit set $\mathscr{K}$. Specifically, we define rescaled position vectors $\mathbf{k}_{n,i}$, $n = 0,1,\dots,$ $i = 0,1,2,3,4,$

$$\begin{aligned}\mathbf{k}_{0,i} &= \lambda^{-1}\mathbf{x}_i(K_{0+}) = \lambda^{-2}\mathbf{x}_i(K_{0-}),\\ \mathbf{k}_{2l-1,i} &= \lambda^{2l-2}\mathbf{x}_i(K_{2l-1}) + (0,1), \quad l = 1,2,\dots\\ \mathbf{k}_{2l,i} &= \lambda^{2l-1}\mathbf{x}_i(K_{2l}), \quad l = 1,2,\dots\end{aligned}$$

and the polygons

$$\mathscr{K}_n = \{\mathbf{k}_{n,0},\dots,\mathbf{k}_{n,|K_n|}\},$$

where $|K_n| = 5 \times 4^{\lceil n/2 \rceil}$.

Note that the vertices of the sequence of $\mathscr{K}_n$ are nested:

$$\mathbf{k}_{n,i} = \mathbf{k}_{n+2m,4^m i}.$$

Moreover, as n tends to infinity, the edges of $\mathscr{K}_n$ scale as τ^{-2n}, while the number of edges increases proportional to 2^{2n}, and so, by the usual arguments, the resulting limit set $\mathscr{K} = \lim_n \mathscr{K}_n$ will have fractal dimension $\log 2/\log \tau$.

Our aim in this section is to relate the asymptotic long-time behavior of round-off orbits to geometrical properties of the fractal $\mathscr{K}$. Specifically, for χ a continuous function of lattice coordinate differences, we will relate moments $\langle\chi\rangle_t$ to *geometrical* moments of $\mathscr{K}$. The analogue of temporal displacement t is an "angle" α representing the fraction of a polygon's perimeter contained between two vertices. This is a quantity which has a chance of remaining meaningful in the limit $n \to \infty$. For the polygon $\mathscr{K}_n$, walking from vertex $\mathbf{k}_{n,i}$ to vertex $\mathbf{k}_{i+j}$ corresponds to an angular displacement

$$\alpha_{n,j} = j/|K_n|,$$

and we define the expectation value of our function χ for j-step displacement vectors as

$$\mu_n(\alpha_{n,j}) = \frac{1}{|K_n|} \sum_{i=0}^{|K_n|-1} \chi(\mathbf{k}_{n,i+\alpha_{n,j}|K_n|} - \mathbf{k}_{n,i}).$$

Due to the odd-even pairing of the $\mathscr{K}_n$ and the $|K_n|$ periodicity, the function μ_n satisfies

$$\mu_{2l-1}(\alpha_{2l-1,j}) = \mu_{2l}(-\alpha_{2l,j}),$$
$$\mu_n(\alpha_{n,j}) = \mu_n(\alpha_{n,j+|K_n|}) = \mu_n(\alpha_{n,j}+1).$$

7.6.4 Substitution rules for lattice walks

The route to asymptotic self-similarity for the sequence of lattice walks generated by periodic orbits of the $p/q = 1/5$ round-off map is not nearly as straightforward as for the Koch construction. The starting point is, of course, to find a substitution rule taking us from one level to the next. Using the vector-field displacement list

$$D_V = ((1,0),(1,1),(0,1),(-1,0),(-1,-1),(0,-1))$$

turns out to be inadequate for this purpose. Instead, we use (Lowenstein and Vivaldi, 1998)

$$D = ((-1,0),(0,-1),(1,0),(0,1),(0,0),(-1,1),(-1,-1),(0,-1),(1,0),(1,1)).$$

The code substitution ϕ_V from $\mathscr{W}(D)$ to $\mathscr{W}(D_V)$ is

$$\phi_V(2) = \phi_V(8) = (0), \qquad \phi_V(9) = (1), \qquad \phi_V(3) = (2),$$
$$\phi_V(0) = (3), \qquad \phi_V(6) = (4) \qquad \phi_V(1) = \phi_V(7) = (5),$$
$$\phi_V(4) = \emptyset, \qquad \phi_V(5) = (3,2).$$

Here the symbol $\emptyset$ denotes the null sequence. With list D, using ϕ_V to translate code sequences into vector-field walks, it is possible to construct the periodic-orbit walks of arbitrary level, starting with

$$W_1 = (\mathbf{d}_0,\mathbf{d}_1,\mathbf{d}_2,\mathbf{d}_3,\mathbf{d}_4), \qquad W_2 = (\mathbf{d}_5,\mathbf{d}_9,\mathbf{d}_8,\mathbf{d}_7,\mathbf{d}_6),$$

and iterating the map $\sigma : \mathscr{W}(D)->\mathscr{W}(D)$ induced by the code substitution ϕ defined by

$$\left.\begin{array}{l} \phi(k) = (k,(k+1)_5, 5+(k+2)_5), \\ \phi(k+5) = (k+5, 5+(k+4)_5,(k+3)_5,(k+4)_5,k+5), \end{array}\right\} \quad k = 0,\ldots,4.$$

One obtains

$$W_{2l-1} = \sigma^{l-1}(W_1), \qquad W_{2l} = \sigma^{l-1}(W_2).$$

From the construction, the numbers of steps of each type at level l are found by applying the $(l-1)$st power of the counting matrix

$$\mathbf{M} = \begin{pmatrix} 1 & 1 & 0 & 0 & 0 & 0 & 0 & 1 & 0 & 0 \\ 0 & 1 & 1 & 0 & 0 & 0 & 0 & 0 & 1 & 0 \\ 0 & 0 & 1 & 1 & 0 & 0 & 0 & 0 & 0 & 1 \\ 0 & 0 & 0 & 1 & 1 & 1 & 0 & 0 & 0 & 0 \\ 1 & 0 & 0 & 0 & 1 & 0 & 1 & 0 & 0 & 0 \\ 0 & 0 & 0 & 1 & 1 & 2 & 0 & 0 & 0 & 1 \\ 1 & 0 & 0 & 0 & 1 & 1 & 2 & 0 & 0 & 0 \\ 1 & 1 & 0 & 0 & 0 & 0 & 1 & 2 & 0 & 0 \\ 0 & 1 & 1 & 0 & 0 & 0 & 0 & 1 & 2 & 0 \\ 0 & 0 & 1 & 1 & 0 & 0 & 0 & 0 & 1 & 2 \end{pmatrix}, \tag{7.14}$$

from the right, on the initial row vectors

$$\mathbf{u}_1 = (1,1,1,1,1,0,0,0,0,0), \qquad \mathbf{u}_2 = (0,0,0,0,0,1,1,1,1,1).$$

Summing over the components gives the lengths (in D steps) of the walks W_n,

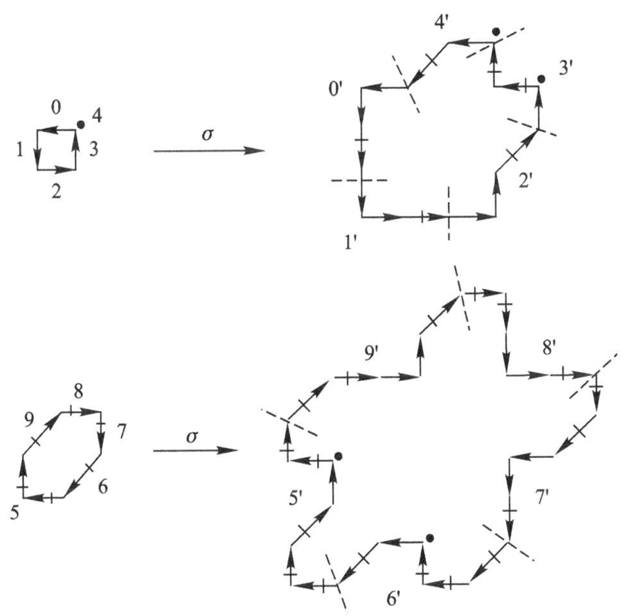

Fig. 7.11 Substitution rules for construction of W_n, from Lowenstein and Vivaldi (1998), Fig. 6.

$$|W_n| = \frac{5}{3}(2^n - (-1)^n), \quad n \geq 1.$$

The corresponding lengths in D_V steps are just the periods $T_j(L)$ in (7.7),written as a single sequence, with $L = \lfloor \frac{n+1}{2} \rfloor$:

$$T_n = \frac{1}{3}(5 \cdot 2^n - 2(-1)^n), \quad n \geq 1.$$

The geometric interpretion of σ in terms of vector-field sequences can be read off from Fig. 7.11.

7.6.5 Separating out an asymptotic walk

We now want to show how iterated application of the substitution ϕ leads asymptotically to fractal snowflakes. Our strategy is to regard ϕ as a linear transformation in a 20-dimensional vector space of displacement-list components, with matrix $\tilde{\mathbf{M}} = \mathbf{M} \otimes \mathbf{I}_2$, with $\mathbf{M}$ given in (7.14) and $\mathbf{I}_2$ equal to the identity matrix in two dimensions. Multiplication by this matrix from the left maps a column of displacements into a column whose entries are the net displacements of the respective substituted code strings.

Eigenvectors of $\tilde{\mathbf{M}}$ correspond to displacement lists with simple scaling properties. The hexagonal Koch construction considered earlier has such a displacement list: writing the components of the $\mathbf{d}_k$, $k = 0,\dots,5$ as a 12-dimensional column vector, multiplication by the analogue of $\tilde{\mathbf{M}}$ reproduces that vector, multiplied by 3. The list D for the $p/q = 1/5$ round-off map does not scale this way, but rather is a linear superposition of the three displacement lists, D_1, D_2, and D_3, shown in Table 7.1. These correspond, respectively to the eigenvalues $\tau^2, 1$, and λ^2 of $\tilde{\mathbf{M}}$. One can verify the following decomposition of D: for $k = 0,\dots,9$,

$$\mathbf{d}_k(D) = \frac{\tau^2}{\sqrt{5}}\mathbf{d}_k(D_1) + \mathbf{d}_k(D_2) + \frac{\lambda^2}{\sqrt{5}}\mathbf{d}_k(D_3).$$

Consider now the three sequences of walks $W_{n,i}$, $n = 1,2,\dots$, $i = 1,2,3$, based on the three displacement lists D_i, but with the same hierarchy of code sequences as W_n based on D. That $W_{n,1}$ dominates in the asymptotic regime is evident from the following lemmas of Lowenstein and Vivaldi (1998).

Lemma 7.2. *The vertices of $W_{n,1}$ form a subsequence of the vertices of a walk on the Koch-type pentagonal snowflake K_n with $5 \cdot 4^{\lceil n/2 \rceil}$ sides.*

Proof. We establish the embedding of vertices, pictured in Fig. 7.12, by showing that there exist substitutions ϕ_{1+} and ϕ_{1-} which induce maps $\sigma_{1\pm}$ from $\mathscr{W}(D)$ to $\mathscr{W}(D_{\pm})$ such that for $l = 1,2,\dots$,

Table 7.1 The displacement lists D_i, $i = 1,2,3$.

k	$d_k(D_1)$	$d_k(D_2)$	$d_k(D_3)$
0	$(0,-1)$	$(-1,1)$	$(0,1)$
1	$(1,0)$	$(-1,-1)$	$(-1,0)$
2	$(\lambda,1)$	$(0,-1)$	$(\tau,-1)$
3	$(-\lambda,\lambda)$	$(1,0)$	$(-\tau,\tau)$
4	$(-1,-\lambda)$	$(1,1)$	$(1,-\tau)$
5	$(0,\tau)$	$(-1,-1)$	$(0,\lambda)$
6	$(-\tau,0)$	$(1,-1)$	$(-\lambda,0)$
7	$(-1,-\tau)$	$(1,1)$	$(1,-\lambda)$
8	$(1,-1)$	$(0,1)$	$(-1,1)$
9	$(\tau,1)$	$(-1,0)$	$(\lambda,-1)$

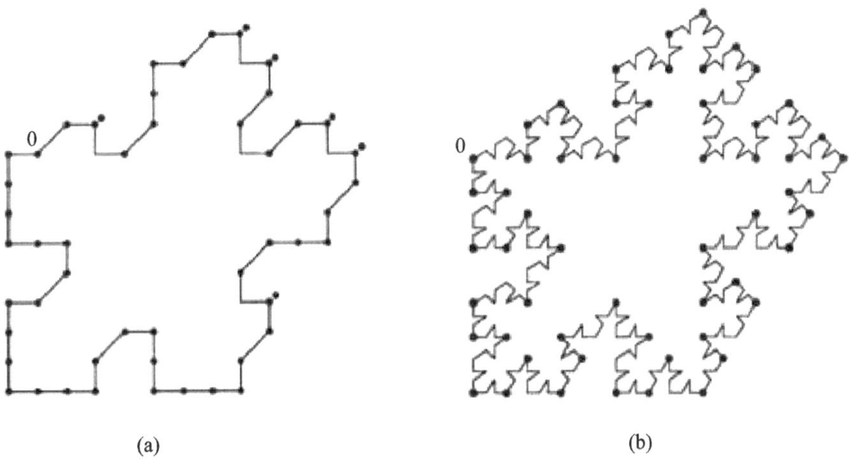

Fig. 7.12 Embedding of $W_{n,1}$ in K_n, from Lowenstein and Vivaldi (1998), Fig. 8.

$$K_{2l-1} = \sigma_{1-}(W_{2l-1}), \qquad K_{2l} = \sigma_{1+}(W_{2l}) \tag{7.15}$$

and, moreover, for $k = 0,\ldots,9$,

$$\sigma_1(d_k(D)) = \sigma_{1+}(d_k(D)) = \sigma_{1-}(d_k(D)). \tag{7.16}$$

Explicitly, the substitutions $\phi_{1\pm}$ are, for $k = 0,\ldots,4$,

$$\phi_{1-}(k) = (k,(k+1)_5,(k+4)_5,k),$$
$$\phi_{1-}(k+5) = ((k+2)_5,(k+3)_5,(k+1)_5,(k+2)_5,(k+3)_5,(k+4)_5,(k+2)_5,(k+3)_5),$$
$$\phi_{1+}(k) = ((k+3)_5,(k+2)_5),$$
$$\phi_{1+}(k+5) = (k,(k+4)_5,(k+1)_5,k).$$

Equation (7.15) is proved by induction. It is easy to verify the equalities for $l = 1$, and then proceed to higher levels using the intertwining relation

$$\sigma_{1\pm}\sigma = \sigma_{\pm}\sigma_{1\pm}.$$

The latter is readily established directly from the explicit definitions of the respective code substitutions.

Lemma 7.3. *$W_{n,2}$ is a walk on the pentagon with vertices*

$$(0,0),(-1,1),(-2,0),(-2,-1),(-1,-1),$$

and so its vertices are bounded in norm by 3.

Proof. The code for $W_{n,2}$ is the same as for W_n, namely $\phi^n(0,1,2,3,4)$ for odd n and $\phi^n(5,9,8,7,6)$ for even n. These code sequences are easily seen to have the simple succession rule: for $k = 0,\dots,4$, the index k is always followed by either $(k+1)_5$ or $5+(k+1)_5$, while $k+5$ is always followed by either $5+(k+4)_5$ or $(k+4)_5$. The reader will readily verify that each step on the pentagon π prescribed by the code is followed by another step on π, either forward or backward. Thus the walk $W_{n,2}$, which necessarily starts with a step from $(0,0)$ to either $(-1,1)$ (n odd) or $(-1,-1)$ (n even), always remains on π. □

Lemma 7.4. *The vertices of $W_{n,3}$, scaled by the factor $\lambda^2/\sqrt{5}$, are bounded in norm by* 1.

Proof. The proof is analogous to that of Lemma 7.2. One shows that the vertices of $W_{n,3}$ are a subsequence of those of a walk $\tilde{K}_n$ generated by a Koch-type construction based once again on the substitutions $\phi_{\pm}$ and initial codes $(0,1,2,3,4)$ and $(0,4,3,2,1)$, but with the displacement lists

$$\begin{aligned}\tilde{D}_- &= ((0,1),(-1,0),(\tau,-1),(-\tau,\tau),(1,-\tau)),\\ \tilde{D}_+ &= \tau\tilde{D}_- = ((0,\tau),(-\tau,0),(\tau^2,-\tau),(-\tau^2,\tau^2),(\tau,-\tau^2)).\end{aligned}$$

The mapping $\tilde{\sigma}_-$, which takes $\tilde{K}_{0-}$ into $\tilde{K}_1$, is illustrated in Fig. 7.13. In contrast to the case of $W_{n,1}$, where the diameter of the polygonal path traversed by the walk increases with n as τ^n, the walks $W_{n,3}$ never leave the confines of a bounded region (the square $[-\tau,\tau]^2$ for n odd, and $[\tau^2,\tau^2]^2$ for n even). Rescaling by a factor $\lambda^2/\sqrt{5}$, we confirm the bound stated in the lemma. For further details of the proof, see Lowenstein and Vivaldi (1998). □

It is now the time to quantify the small differences among the four families of periodic walks we have introduced, namely the vector-field walks $\{V_n\}$, the scaling family $\{W_n\}$, the dominant parts $\{W_{n,1}\}$, and the pentagonal Koch snowflakes $\{K_n\}$. Omitting the tedious details, we state the key lemma bounding the difference between corresponding vertices $\mathbf{x}_t(V_n)$ and $\mathbf{x}_j(K_n)$ (Lowenstein and Vivaldi, 1998).

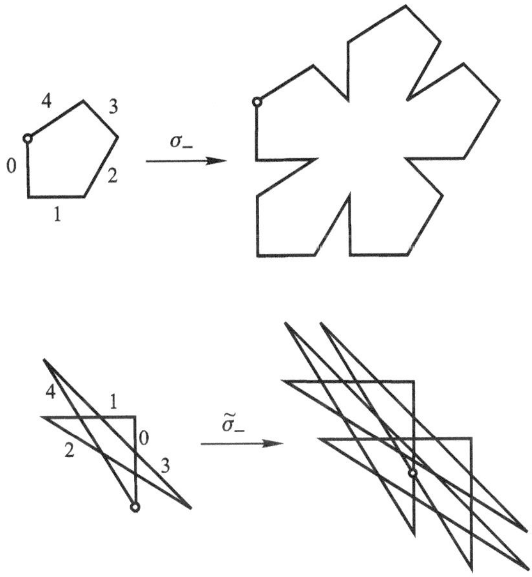

Fig. 7.13 Substitution rules for construction of K_n and $\tilde{K}_n$, from Lowenstein and Vivaldi (1998), Fig. 7.

Lemma 7.5. *For all positive* t, n, *there exist constants* ε_i, $i = 1, 2, 3$, *such that*

$$||\mathbf{x}_t(V_n) - \frac{\tau^2}{\sqrt{5}}\mathbf{x}_{s(n)t}(K_n)|| \leq \varepsilon_1(2^{-n}t)^{1/\delta} + \varepsilon_2\lceil n/2 \rceil + \varepsilon_3,$$

where $s(2l-1) = 6$, $s(2l) = 3$.

7.6.6 Asymptotic scaling

The final steps in deriving the long-time behavior of moments involve a number of of technical lemmas (Lowenstein and Vivaldi, 1998). We will be satisfied here with a sketch of the main arguments and a statement of the final result.

We consider the expectation value $\langle \chi \rangle_t$ of a continuous and piecewise differentiable function (random variable) $\chi(\mathbf{x})$ on $\mathbb{Z}^2$ which is homogeneous of degree $\kappa > 0$, i.e. for all positive ρ,

$$\chi(\rho\mathbf{x}) = \rho^\kappa \chi(\mathbf{x}).$$

The expectation value $\langle \chi \rangle_t$ with respect to $P(\mathbf{x}, t)$ can be decomposed into contributions $\langle \chi \rangle_{n,t}$ from walks over the respective Π_n using (7.11):

$$\langle \chi \rangle_t = \sum_{n=0}^{2L_*(t)} b_n(t) \langle \chi \rangle_{n,t}.$$

Here a walk on Π_n is extended by periodicity to time t if the latter exceeds the return time T_n. Using Lemma 7.5, the partial expectation value $\langle \chi \rangle_{n,t}$ can be well approximated for large t by the corresponding rescaled expectation value on the Koch snowflake,

$$(\lambda^{n+1}\sqrt{5})^{-\kappa} \mu_n(\alpha_{n,s(n)t}),$$

with error terms given by the lemma. Further approximations, with additional error estimates, include: (1) replacing T_n by $\frac{5}{3}2^n$ in the definition of $b_n(t)$; (2) deleting from the sum over n the terms with $n < L_*(t)$; (3) relacement of $\mu_n(\alpha_{n,st})$ by (see Lowenstein and Vivaldi, 1998, Lemma 5.1, for justification)

$$\mu((-1)^n \alpha_{n,st}) = \lim_{p\to\infty} \mu_{n+2p}((-1)^n \alpha_{n,st});$$

(4) making the substitution

$$(-1)^n \alpha_{n,st} = (-2)^{2L_*(t)-n} \alpha_t, \qquad \alpha_t = \frac{3t}{5 \cdot 4^{\lceil \log_4(9t/10) \rceil}} = \frac{1}{6} 4^{\{9t/10\}},$$

where $\{\cdot\}$ is the fractional part function; (5) changing the summation variable to $m = 2L_8(t) - n$; and (6) extending the summation range to infinity. The final result of all these manipulations, accompanied by careful error accounting, is the following theorem (Lowenstein and Vivaldi, 1998, Theorem 5.2).

Theorem 7.3. *Let χ be a homogeneous random variable of degree $\kappa > 2 - \delta$, where $\delta = \log 2/\log \tau$. Then the following holds:*

$$\langle \chi \rangle_t = t^{\beta}(G(t) + O(t^{-a} \log^b t)), \tag{7.17}$$

where

$$\beta = 1 + \frac{\kappa - 2}{\delta}, \qquad a = \min\left\{\frac{\beta}{2}, \frac{1}{\delta}\right\}, \qquad b = 1 + \frac{1}{\delta},$$

$$G(t) = F(\alpha_t),$$

$$F(\alpha) = \frac{1}{6}\left(\frac{3}{5\alpha}\right)^{\beta} (\lambda\sqrt{5})^{2-\kappa} \sum_{m=0}^{\infty} \frac{c_m}{2^{\beta m}} \mu((-2)^m \alpha),$$

with $c_0 = 1, c_1 = \lambda$, and $c_m = \sqrt{5}\lambda^4$ for $m > 1$. The form factor $G(t)$ is log-periodic, i.e. is a function of $\log t$ *with period* $\log 4$.

We thus have, for the moments of the probability distribution, asymptotic power-law t^{β} modulated by a log-periodic form factor $G(t)$. If we think of these formulas as describing some sort of diffusive propagation of the round-off error, the diffusion is decidedly anomalous, even in the special case of the second x-moment $\langle x^2 \rangle_t$: the second transport coefficient, defined as

$$g_2(t) = t^{-1}(\langle x^2 \rangle_t - \langle x \rangle_t^2),$$

is just $G(t)$, which fails to tend toward a limit for t tending to infinity, in contrast to the convergent transport coefficient of normal diffusion.

Lowenstein and Vivaldi (1998) made a high-resolution numerical calculation of the form factor $G(t)$ for $\chi(m,n) = m^2$, with t in the range $683 \leq t \leq 2730$. Their efficient algorithm exploited a linear recursion relation for the 2-point correlation function of the snowflakes K_n. The plot is shown in Fig. 7.14.

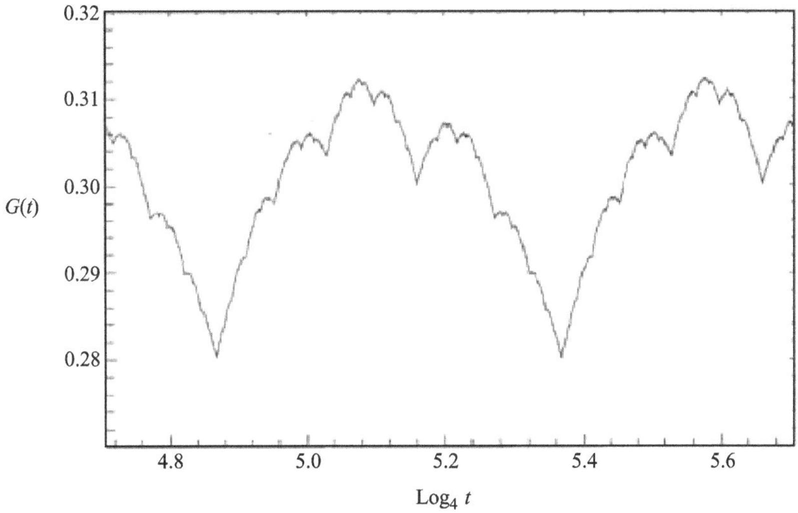

Fig. 7.14 Plot of the form factor $G(t)$ for $683 \leq t \leq 2730$, from Lowenstein and Vivaldi (1998), Fig. 9.

7.7 Model I

We now explore the asymptotic long-time behavior of the probability distribution $P(\mathbf{x},t)$ and its moments using the computer assisted methods of Chapter 6. For the sake of comparison, we will consider precisely the same local maps which we emphasized in the context of kicked oscillator maps, namely Models I and II, with $p/q = 3/10$ and $5/12$. The underlying symbolic dynamics is, of course, exactly the same, while the lattice isometries driven by the symbolic machine are completely different. In particular, the lattice rotation number is always $1/4$ for our kicked oscillators, while for the corresponding round-off model it has the same denominator as that of the local map.

We treat first Model I, using the same list of 501 time intervals and level-8 tiling of the local square that we employed in Chap. 6. The list of participating domains (the three scaling domains $D_j(8)$ with nonzero area, plus 17 periodic domains) are listed with their return times and areas in Table 6.1. Choosing a representative lattice point localized in each of the 20 domains, we calculated and stored the orbits obtained by iterating the round-off map over a complete return time, augmented by $t_*(8) = 218450$. The time-t lattice displacements, for each of the 501 t values and each orbit point were calculated and their weighted contribution to $P((m,n),t)$ and to the expectation values of $|m|, m^2$, and m^4, accumulated. The results are displayed in Figs. 7.15 and 7.16.

As we might have anticipated from Sect. 7.6.6, the kth moments are characterized asymptotically by power-law growth modulated by log-periodic form factors:

$$\langle |x|^k \rangle_t \sim t^{1+\frac{k-2}{\delta}} G_k(t)$$

where δ is the fractal dimension associated with the local-map scaling (ratio of the logarithms of the temporal and spatial scale factors), as well as the asymptotic

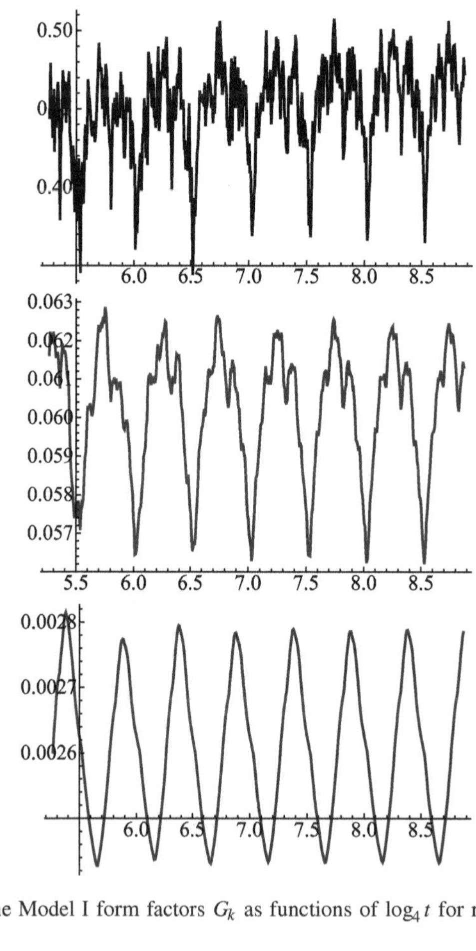

Fig. 7.15 Plots of the Model I form factors G_k as functions of $\log_4 t$ for moments $\langle |m| \rangle_t$, $\langle m^2 \rangle_t$, and $\langle m^4 \rangle_t$.

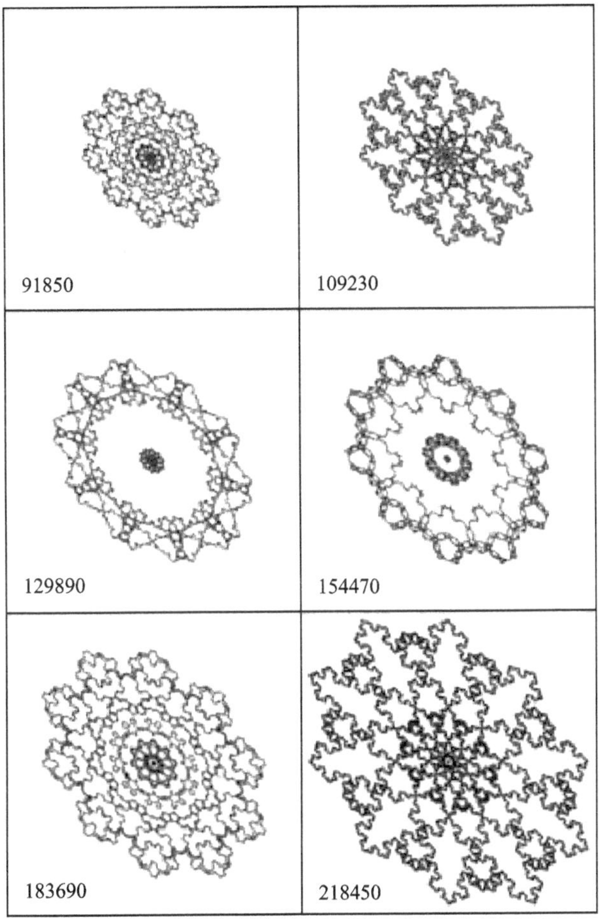

Fig. 7.16 Evolution of the Model I probability distribution. Successive t values differ by approximately a factor of $2^{1/4}$.

scaling of the periodic orbits on the lattice, and the asymptotic logarithmic period of G_k is one-half the logarithm of the temporal scale factor. From Fig. 7.16, we see that the support of the probability distribution also exhibits apparent power-law growth with log-periodic modulation.

7.8 Model II

Perhaps the most fascinating example of long-time asymptotics among the class of renormalizable round-off models is Model II, with rotation number $5/12$, thanks to the presence of two competing pseudochaotic webs with different dimensions. From the preceding sections, we expect the moments to behave as

$$\langle |x|^k \rangle_t \sim t^{1+\frac{k-2}{\delta_A}} G_k^A(t) + t^{1+\frac{k-2}{\delta_B}} G_k^B(t), \tag{7.18}$$

where

$$\delta_A = -\frac{\log 4}{\log(2-\sqrt{3})} = 1.05\ldots, \qquad \delta_B = -\frac{\log 5}{\log(2-\sqrt{3})} = 1.22\ldots,$$

This means that, for $k < 2$, the B term will dominate for asymptotically large t, while the A term will be dominant for $k > 2$. The especially interesting case is $k = 2$, since there the dependence on dimension drops out, and (7.18) reduces to

$$\langle |x|^k \rangle_t \sim t \left(G_k^A(t) + G_k^B(t) \right).$$

The two form factors oscillate with incommensurate frequencies, and hence the quasi-diffusive power law is modulated by a *log-quasiperiodic* factor.

The expected long-time behavior in Model II has been verified by Lowenstein and Liu (2003) in a numerical calculation which uses essentially the method applied to Model I in the preceding section. The authors used a list of 441 time intervals approximately equally spaced on a logarithmic scale between 21000 and 24982500 for the two sequences of periodic orbits in the A hierarchy, with scaling limit point $(41+72\tau, 41+72\tau)$, and 292 t values between 21000 and 2271000 for sequences in the B hierarchy, with limit point (τ, τ), where $\tau = -1/\sqrt{3}$. The log-log plots of $\langle x^2 \rangle_t$ and $\langle x^2 \rangle_t / t$ in Fig. 7.17 provide strong evidence for the validity of the asymptotic logarithmic quasiperiodicity.

In the absence of a rigorous asymptotic analysis analogous to that of Sect. 7.6.6, it is important to probe more deeply the asymptotic power laws embodied in (7.18), making evident the remarkable competition between the two scaling hierarchies with different fractal dimensions. The numerical experiment described below fills in some of the missing pieces. It also provides a fitting way to wrap up the final chapter of this book, since it exploits in a powerful and practical way its major themes of recursive tiling and underlying symbolic dynamics.

The new investigation extends the results of Lowenstein and Liu (2003) in several ways. First of all, the calculation treats time intervals uniformly for the two scaling hierarchies and the nonscaling orbits as well, in the range $7200000 \leq t \leq 36000000$, while the previous work assumed maximum times 2271000 and 24982500 for hierarchies A and B, respectively, and omitted the nonscaling orbits from the probability sums. While the latter are bounded on the lattice give no significant contribution to the asymptotics of the second moment, they do provide an important component of the probability distribution for small excursions on the lattice. A further innovation in the current experiment is the exploration of moments $\langle (|m| + |n|)^k \rangle$ over a sufficiently wide range of k values to test the asymptotic behavior conjectured in (7.18), including the alternating dominance of the A and B hierarchies for k greater than or less than 2, respectively. Finally, the computation exploits fully the symbolic dynamics underlying the propagation of round-off error on the lattice: each map iteration is executed by updating a finite codometer.

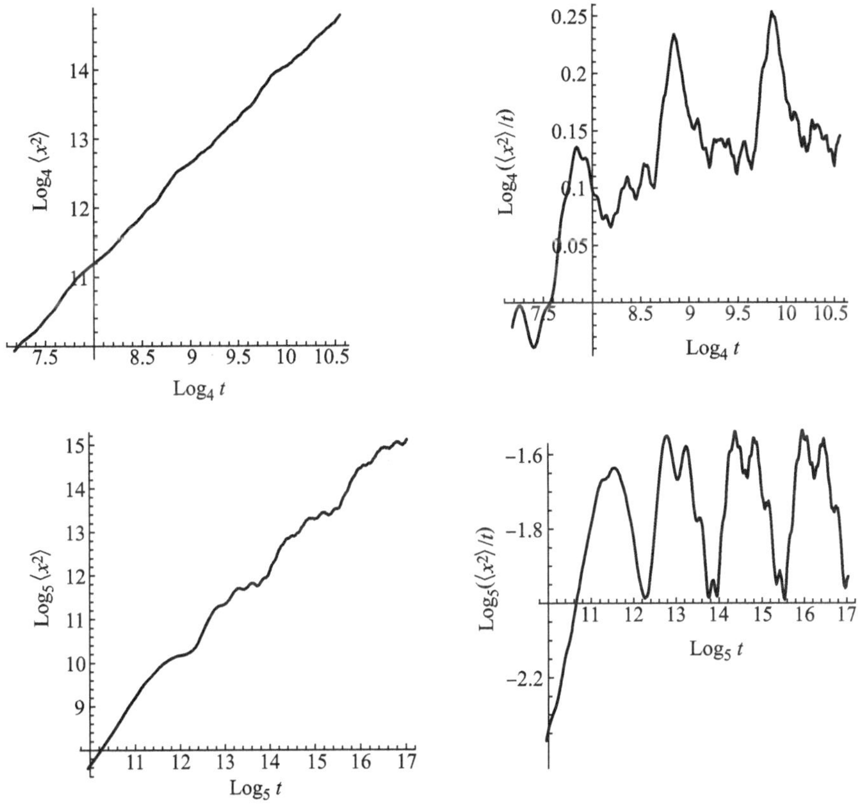

Fig. 7.17 Log-log plots (Lowenstein and Liu, 2003, Figs. 17, 20–22) of $\langle x^2 \rangle$ vs. t (on the left) and the corresponding form factors (on the right) for the scaling sequences of the $5/12$ model associated with the limit points (τ, τ) (above) and $(41+72\tau, 41+72\tau)$ (below).

In Model II, the recursive tiling property (see Sect. 2.3) allows us to partition the square Ω, apart from a set of measure zero, into three disjoint families of convex polygons: Family A, consisting of domains associated with the scaling limit point (τ,τ), $\tau = -\sqrt{3}/3$, Family B, consisting of domains associated with the second scaling limit, $(41+72\tau, 41+72\tau)$, and Family C, consisting of periodic domains not associated with either scaling limit. For arbitrary (non-negative integer) choice of level $L^{(\mathrm{A})}$ (resp. $L^{(\mathrm{B})}$) in the scaling hierarchy, Family A will consist of the domains $D_j(L^{\mathrm{A}})$, $j = 0,\ldots,J$, together with all domains which are periodic under the return maps of levels less than $L^{(\mathrm{A})}$, with an analogous property for Family B. The choices of $L^{(\mathrm{A})}$ and $L^{(\mathrm{B})}$ are constrained by the selection of time interval t in the construction of $P(\mathbf{r},t)$. For times in our target range, since $t_*^{(\mathrm{A})}(6), t_*^{(\mathrm{B})}(5) > 36000000$, it is sufficient to choose

$$L^{(\mathrm{A})} = 6, \qquad L^{(\mathrm{B})} = 5.$$

The probability distribution then has the expansion

$$P(\mathbf{r},t) = \tau^{-2} \sum_{D} \sum_{s=0}^{T(D)-1} \mathscr{A}(D) \delta(\Phi^{s+t}(\mathbf{r}_D) - \Phi^{s}(\mathbf{r}_D), \mathbf{r}),$$

where $T(D)$ is the return time of domain D under iteration of Φ, and D runs over the 93 domains in the following list (the chosen tiling of Ω):

$$\begin{array}{ll} D_j^{(\mathrm{A})}(6), & j = 0,2,4,7,10,19,21,22,12,14,15,17, \\ \Pi_j^{(\mathrm{A})}(L), & L = 0,\ldots,5, \quad j = 0,\ldots,7, \\ D_j^{(\mathrm{B})}(5), & j = 0,3,5,6,8,10, \\ \Pi_j^{(\mathrm{B})}(L), & L = 0,\ldots,4, \quad j = 0,1, \\ \Pi_j^{\mathrm{misc}}, & j = 0,\ldots,16. \end{array}$$

The relatively short periodic orbits, Π_j^{misc} , were handled as in Sect. 7.7. Using a modified bisection method, a single point with algebraic-integer coordinates was found in each of the initial domains, and the corresponding lattice points were calculated using $\mathscr{E}^{-1}$. Here we are relying on the fact that all the orbits starting at such points have the same shape, hence lead to the same probability distribution for displacements. For each initial point, two lattice orbits, displaced in time by a specific time interval t, were generated by explicit iteration of the map Φ using exact, computer assisted algebra. The area-weighted contributions to $P(\mathbf{x},t)$ and $\langle(|x|+|y|)^k\rangle$ were recorded at each step.

For the remaining initial domains, a more efficient approach, based on symbolic dynamics, was employed. To illustrate the method, let us consider a specific domain D in the tiling, with return time T, and describe the determination of D's contribution to the second moment $\langle(|m|+|n|)^2\rangle$. For simplicity, we assume D is a periodic domain $\Pi_j^{(\mathrm{A})}(L)$; the treatment of the others follows the same pattern.

The dynamical engine of our numerical calculation is a finite Vershik codometer with registers (i_a, j_a, t_a), $a = 0,\ldots,L+1$. Thanks to scaling, registers 1 through 5 operate identically, with the predecessor function $p(i,j,t)$ corresponding to the $\rho^{(\mathrm{A})}(L-1)$ return path of the domain $D_j^{(\mathrm{A})}(L)$. Register 0 keeps track of individual Φ iterations, with j_0 labeling return orbits of the 12 domains $D_{j_0}^{(\mathrm{A})}(0)$, while register 6 keeps track of the domains $D_{j_L}(L)$ visited by D in its return orbit. The initial codometer setting is

$$\left(\begin{pmatrix} 0 \\ 22 \\ 0 \end{pmatrix} \begin{pmatrix} 1 \\ 22 \\ 0 \end{pmatrix} \begin{pmatrix} 1 \\ 22 \\ 0 \end{pmatrix} \begin{pmatrix} 1 \\ 22 \\ 0 \end{pmatrix} \begin{pmatrix} 1 \\ 22 \\ 0 \end{pmatrix} \begin{pmatrix} 1 \\ 17 \\ 0 \end{pmatrix} \begin{pmatrix} 2 \\ 7 \\ 0 \end{pmatrix} \right)$$

where we have assigned indices $i = 0,1,2$ to label the three distinct register species. After T time steps, the codometer resets automatically to its initial sequence.

At each time step, the state of register 0 is used to update an array storing the coordinates of 12 points, (x_b, x_b), $b = 0,\ldots,11$, one on each branch of the lattice orbit. Since the probability and moment calculations involve only vector displacements between orbit points on the same branch, all 12 points of the lattice array were conveniently set to $(0,0)$ at the outset.

Our numerical experiments measured three sets of data:

Moments

The moments $\langle(|x|+|y|)^{k/4}\rangle$, $k=5,\ldots,16$ are area-weighted average over all domains of the recursive tiling of the square. All vector dispacements were calculated exactly and the converted to floating point approximate real numbers before summation. This avoided the need for arbitrary precision arithmetic, but introduced an acceptable numerical error in the summation. The case $k=0$ (total area) was included as a check on accuracy and completeness. The results are displayed in Fig. 7.18 and Table 7.2. In the latter, we measured the scaling exponents associated with times which differ by a factor 4 in case A and by a factor 5 in case B and compared them with the predicted asymptotic values from (7.18). Apart from the case $k=1$, the agreement is within a few percent, improving with increasing k. It seems reasonable to assume that increasing the range of t would improve the convergence considerably.

Second moment for $7200000 \leq t \leq 36000000$

The second moment was calculated for 401 points, equally spaced on a logarithmic scale, over the range $7200000 \leq t \leq 36000000$. Log-log plots of the second moment and form factor, for both the A and B scaling sequences, are shown in Fig. 7.19. This extends to a longer time interval, with greater detail, the results of Lowenstein and Liu (2003), shown in Fig. 7.17. The results confirm the conjectured quasiperiodicity of the form factor of the second moment.

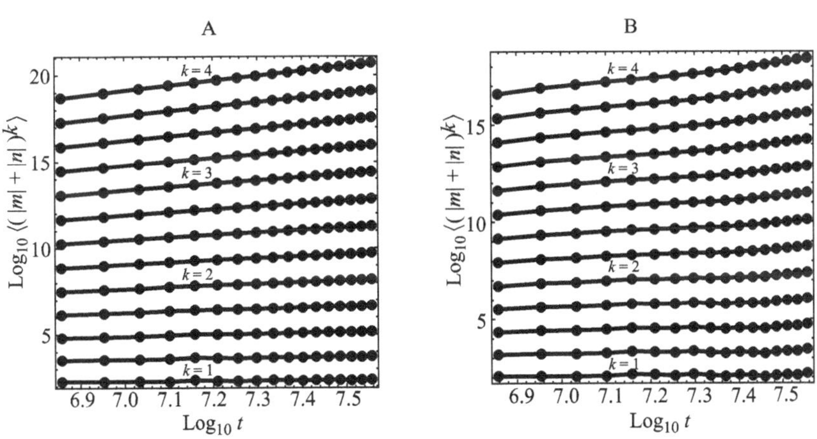

Fig. 7.18 Log-log plot of the moments $\langle(|x|+|y|)^k\rangle_t$, for 17 equally spaced time intervals t in the range $7200000 \leq t \leq 36000000$ and $k=1,1.25,1.5,\ldots,4$.

Probability distributions

The probability distributions $P^{(f)}(\mathbf{x},t)$, $f = \mathrm{A,B,C}$, coarse grained to display the results at four different scales, were calculated for 17 equally spaced t values in the range $7200000 \leq t \leq 36000000$. The combined probability for two different scales at $t = 36000000$ are shown in Figs. 7.20 and 7.21.

Table 7.2 Scaling exponents $\beta^{(\mathrm{A})}$ and $\beta^{(\mathrm{B})}$ from the data. The measured values of $\beta^{(\mathrm{A})}$ in the second column are taken from the moments for $t = 9000000$ and $t = 36000000$; those of $\beta^{(\mathrm{B})}$ in the fourth column use the moments for $t = 7200000$ and $t = 36000000$. The third and fifth columns give the predicted values of the exponents, from (7.18).

k	$\beta^{(\mathrm{A})}$	$1+(k-2)/\delta_{\mathrm{A}}$	$\beta^{(\mathrm{B})}$	$1+(k-2)/\delta_{\mathrm{B}}$
1	0.161035	0.0500157	0.225791	0.181728
1.25	0.325134	0.287512	0.397758	0.386296
1.5	0.530483	0.525008	0.591461	0.590864
1.75	0.755445	0.762504	0.793556	0.795432
2	0.988014	1	0.998148	1
2.25	1.22331	1.2375	1.20318	1.20457
2.5	1.45962	1.47499	1.40808	1.40914
2.75	1.69634	1.71249	1.61278	1.6137
3	1.93326	1.94998	1.81729	1.81827
3.25	2.17031	2.18748	2.0217	2.02284
3.5	2.40743	2.42498	2.22604	2.22741
3.75	2.64462	2.66247	2.43035	2.43198
4	2.88186	2.89997	2.63464	2.63654

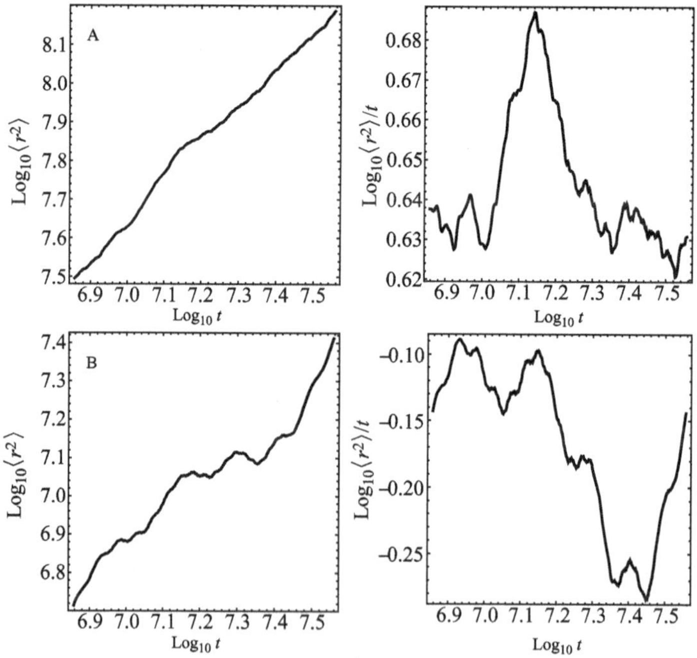

Fig. 7.19 Log-log plots of the second moment $\langle r^2\rangle$, $r = |x| + |y|$ and form factor $\langle r^2\rangle/t$, for the A and B scaling sequences

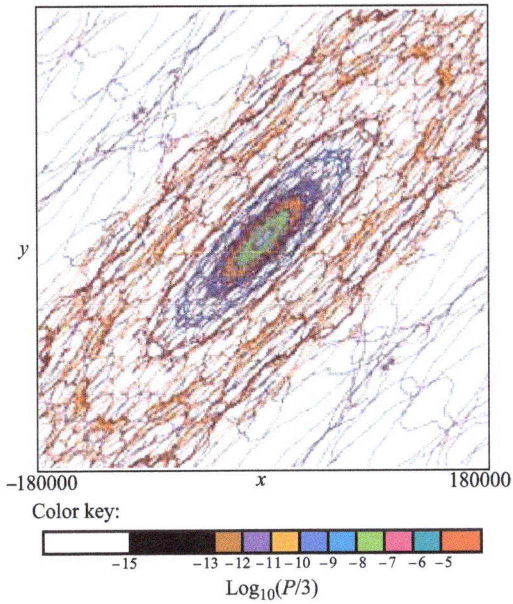

Fig. 7.20 Coarse-grained probability distribution for the round-off error for a time interval $t =$ 36000000. There are 600×600 pixels, each covering 600^2 lattice sites. Colors are assigned to the indicated ranges of the total probability P per pixel. (See the color figure at the end of the book.)

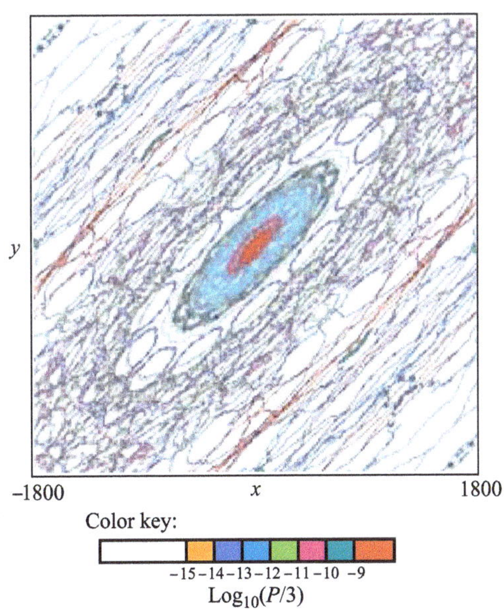

Fig. 7.21 Coarse-grained probability distribution for the round-off error for a time interval $t =$ 36000000. There are 600×600 pixels, each covering 6^2 lattice sites. Colors are assigned to the indicated ranges of the total probability P per pixel. (See the color figure at the end of the book.)

7.9 A conjecture

Based on the analytical and numerical results of this chapter, we can feel reasonably confident that we understand the asymptotic long-time behavior of the renormalizable round-off models. Only the case of rotation number $1/5$ was treated with mathematical rigor, but the picture was sufficiently clear that we are able to formulate a conjecture to cover all of the relevant models. For these, every point of the $\mathbb{Z}^2$ lattice is localized, via the map $\mathscr{E}$, in a polygon belonging to one of a finite number of scaling sequences, or else to one of a finite number of periodic polygons. Each scaling sequence is associated with an infinite sequence of lattice walks which tend in the limit to fractal snowflakes. Moreover each scaling sequence has a unique fractal dimension δ, and contributes, for time t tending to infinity, to the expectation value of a homogeneous random variable χ of degree κ as

$$\langle\chi\rangle_t = t^{\beta}(G_{\kappa}(t) + O(t^{-a}\log^b t)), \qquad \beta = 1 + \frac{\kappa - 2}{\delta},$$

where a and b are positive, model dependent, constants and the form factor G_{κ} is a periodic function of $\log t$ which can be expressed in terms of the expectation value of χ over the corresponding fractal snowflake. Our numerical experiments in Models I and II provide strong evidence for the validity of the conjectured modulated power law.

References

Akiyama S., Brunotte H., Pethö A., and Steiner W. (2008) Periodicity of certain piecewise affine planar maps, *Tsukuba Journal of Mathematics* **32**, 1–55.

Hardy G. H. and Wright E. M. (1979) *An Introduction to the Theory of Numbers*, 5th edition, Oxford University Press, Oxford.

Kouptsov K.L., Lowenstein J.H., and Vivaldi F. (2002) Quadratic rational rotations of the torus and dual lattice maps, *Nonlinearity* **15**, 1795–1842.

Lowenstein J.H., Hatjispyros S., and Vivaldi F. (1997) Quasi-periodicity, global stability and scaling in a model of Hamiltonian round-off, *Chaos* **7**, 49–66.

Lowenstein J.H. and Liu Sangtian (2003) Propagation of round-off error in a model of quadratic rational rotations, *Communications in Nonlinear Science and Numerical Simulation* **8**, 215–237.

Lowenstein J.H. and Vivaldi F. (1998) Anomalous transport in a model of Hamiltonian roundoff, *Nonlinearity* **11**, 1321–1350.

Lowenstein J.H. and Vivaldi F. (2000) Embedding dynamics for round-off errors near a periodic orbit, *Chaos* **10**, 747–755.

Vivaldi, F. (1994) Periodicity and transport from round-off errors, *Experimental Mathematics* **3**, 303–315.

Appendix A
Data Tables[1]

A.1 Model I Data Tables, from Kouptsov et al. (2002)

A.1.1 Generating domain

Below is a table of generating domains and subdomains, and the base domain for the scaling sequence. The coordinates for the base domain are relative to the scaling limit point $\mathbf{x}_\infty$. #V is the number of vertices of a domain (see Fig. 2.8). The map K acts on the generating subdomain Ω_j as $\mathbf{x} \mapsto \mathbf{C}\cdot\mathbf{x} + \mathbf{d}_j(\Omega)$.

Domain	#V	$\mathscr{D}(\mathbf{m},\mathbf{s},\mathbf{b})$	$\mathbf{d}_j(\Omega)$
Ω	4	$\mathscr{D}((0,1,0,1),(1,1,-1,-1),(0,0,\tau,\tau))$	—
Ω_0	3	$\mathscr{D}((1,5,7),(-1,-1,-1),(\tau,0,-\tau))$	$(0,-\tau)$
Ω_1	5	$\mathscr{D}((0,1,2,5,7),(-1,-1,-1,-1,-1),(\tau,\tau,\tau,0,0))$	$(0,0)$
Ω_2	3	$\mathscr{D}((0,2,6),(-1,-1,-1),(\tau,0,0))$	$(0,\tau)$
Ω_3	2	$\mathscr{D}((0,2,5,7),(-1,1,-1,1),(\tau,0,0,0))$	$(0,0)$
Ω_4	2	$\mathscr{D}((0,2,5,7),(-1,1,-1,1),(-1,\tau,0,-\tau))$	$(0,-\tau)$
Ω_5	2	$\mathscr{D}((0,1,5,6),(-1,1,-1,1),(\tau,0,0,0))$	$(0,\tau)$
Ω_6	2	$\mathscr{D}((1,0,6,5),(-1,1,-1,1),(-1,0,0,0))$	$(0,0)$
Ω_7	2	$\mathscr{D}((1,0,6,5),(-1,1,-1,1),(\tau,0,1,0))$	$(0,-\tau)$
Ω_8	1	$\mathscr{D}((0,2,5,7),(1,1,1,1),(0,0,0,0))$	$(0,0)$
Ω_9	1	$\mathscr{D}((0,2,5,7),(1,1,1,1),(0,\tau,0,-\tau))$	$(0,-\tau)$
$D(0)$	3	$\mathscr{D}((5,1,7),(-1,-1,-1),(0,0,-1))$	—

A.1.2 Level-0 scaling sequence domains

j	#V	$\mathscr{D}(\mathbf{m},\mathbf{s},\mathbf{b})$	$T_j(0)$	$\mathbf{d}_j(0)$
0	4	$\mathscr{D}((1,3,5,7),(-1,-1,-1,-1),(0,-\alpha_{1,1},0,-\alpha_{1,0}))$	6	$(0,\alpha_{1,0})$
1	2	$\mathscr{D}((0,3,5,8),(-1,1,-1,1),(-\alpha_{2,1},-\alpha_{1,1},0,\alpha_{1,1}))$	19	$(0,\alpha_{2,1})$
2	3	$\mathscr{D}((1,5,7),(-1,-1,-1),(0,0,-\alpha_{2,1}))$	10	$(\alpha_{1,1},0)$
3	2	$\mathscr{D}((0,2,5,7),(-1,1,-1,1),(-\alpha_{2,1},\alpha_{2,1},0,-\alpha_{2,1}))$	27	$(\alpha_{3,2},-\alpha_{1,1})$
4	3	$\mathscr{D}((2,5,8),(-1,-1,-1),(\alpha_{2,1},0,\alpha_{1,1}))$	18	$(0,-\alpha_{1,1})$

[1] In these tables, we use the abbreviations $\alpha_{r,s} = r + s\tau$ and $\alpha^L_{r,s} = \omega^{L+1}(r+s\tau)$.

A.1.3 Level-0 periodic domains

j	#V	$\mathscr{D}(\mathbf{m},\mathbf{s},\mathbf{b})$	$\tilde{T}_j(0)$
0	2	$\mathscr{D}((2,1,7,6),(1,-1,1,-1),(\alpha_{2,1},\alpha_{5,3},-\alpha_{2,1},\alpha_{3,2}))$	25
1	1	$\mathscr{D}((2,1,7,6),(1,1,1,1),(\alpha_{2,1},\alpha_{5,3},-\alpha_{2,1},-\alpha_{5,3}))$	33
2	2	$\mathscr{D}((2,1,7,6),(1,-1,1,-1),(\alpha_{2,1},-\alpha_{8,5},-\alpha_{2,1},-\alpha_{5,3}))$	105
3	1	$\mathscr{D}((2,1,7,6),(1,1,1,1),(\alpha_{2,1},-\alpha_{8,5},-\alpha_{2,1},\alpha_{8,5}))$	57
4	5	$\mathscr{D}((0,2,4,6,8),(-1,-1,-1,-1,-1),(\alpha_{1,1},\alpha_{4,2},\alpha_{1,0},\alpha_{3,2},\alpha_{2,2}))$	24
5	5	$\mathscr{D}((1,3,5,7,9),(-1,-1,-1,-1,-1),(0,-\alpha_{1,1},0,-\alpha_{1,0},-\alpha_{1,0}))$	6

A.1.4 Miscellaneous periodic domains

j	#V	$\mathscr{D}(\mathbf{m},\mathbf{s},\mathbf{b})$	$\tilde{T}_j(\Omega)$
0	1	$\mathscr{D}((0,1,5,6),(1,1,1,1),(0,-2-\tau,0,2+\tau))$	13
1	2	$\mathscr{D}((0,1,5,6),(1,-1,1,-1),(0,-2-\tau,0-3-2\tau))$	25
2	1	$\mathscr{D}((0,1,5,6),(1,1,1,1),(0,3+2\tau,0,-3-2\tau))$	17
3	1	$\mathscr{D}((0,1,5,6),(1,1,1,1),(0,1+\tau,0,-1-\tau))$	7
4	1	$\mathscr{D}((0,1,5,6),(1,1,1,1),(0,0,0,0))$	1
5	2	$\mathscr{D}((0,1,5,6),(1,-1,1,-1),(0,-1,0,-1-\tau))$	5
6	5	$\mathscr{D}((0,2,4,6,8),(-1,-1,-1,-1,-1),(\tau,1+\tau,1,1,1+\tau))$	4

A.2 Model II Data Tables, from Kouptsov et al. (2002)

A.2.1 Generating partition

Below is a table of generating domains and subdomains, and the base domains for the scaling sequences. The coordinates for the base domain are relative to the respective scaling limit points.

Domain	#V	$\mathscr{D}(\mathbf{m},\mathbf{s},\mathbf{b})$	$\mathbf{d}_j(\Omega)$
Ω	4	$\mathscr{D}((0,1,0,1),(1,1,-1,-1),(0,0,\tau,\tau))$	—
Ω_0	3	$\mathscr{D}((0,1,2),(-1,1,-1),(\tau,0,0))$	$(0,\tau)$
Ω_1	5	$\mathscr{D}((0,2,0,2,1),(-1,1,1,-1,-1),(\tau,0,0,\tau,\tau))$	$(0,0)$
Ω_2	3	$\mathscr{D}((0,1,2),(1,-1,1),(0,\tau,\tau))$	$(0,-\tau)$
$D^{(A)}(0)$	3	$\mathscr{D}((4,11,7),(-1,-1,-1),(0,0,209+362\tau))$	—
$D^{(B)}(0)$	3	$\mathscr{D}((11,3,6),(-1,-1,-1),(0,209+362\tau,0))$	—

A.2.2 *Level-0 scaling domains, sequence A*

j	#V	$\mathscr{D}(\mathbf{m},\mathbf{s},\mathbf{b})$	$\mathbf{d}_j(0)$	$T_j(0)$
0	3	$\mathscr{D}((0,7,4),(-1,-1,-1),(\alpha_{571,989}),(\alpha_{209,362},0))$	$(-\alpha_{209,362},\alpha_{362,627})$	6773
1	2	$\mathscr{D}((11,0,4,6),(-1,1,-1,1),(-\alpha_{780,1351},\alpha_{571,989},0,-\alpha_{571,989}))$	$(0,\alpha_{571,989})$	20187
2	4	$\mathscr{D}((5,7,11,2),(-1,-1,-1,-1),(\alpha_{780,1351},\alpha_{209,362},0,\alpha_{571,989}))$	$(-\alpha_{209,362},\alpha_{571,989})$	13546
3	2	$\mathscr{D}((6,5,2,11),(-1,1,-1,1),(-\alpha_{571,989},\alpha_{780,1351},\alpha_{571,989},-\alpha_{780,1351}))$	$(0,\alpha_{571,989})$	20187
4	3	$\mathscr{D}((6,11,2),(-1,-1,-1),(-\alpha_{571,989},-\alpha_{780,1351},\alpha_{571,989}))$	$(0,\alpha_{571,989})$	20319
5	2	$\mathscr{D}((6,2,11,8),(-1,1,-1,1),(-\alpha_{571,989},\alpha_{571,989},0,-\alpha_{571,989}))$	$(-\alpha_{989,1713},\alpha_{1142,1978})$	26647
6	2	$\mathscr{D}((9,1,4,7),(-1,1,-1,1),(\alpha_{780,1351},-\alpha_{780,1351},0,\alpha_{780,1351}))$	$(-\alpha_{780,1351},\alpha_{1351,2340})$	26801
7	3	$\mathscr{D}((4,9,1),(-1,-1,-1),(0,\alpha_{780,1351},-\alpha_{780,1351}))$	$(-\alpha_{780,1351},\alpha_{1351,2340})$	26933
8	2	$\mathscr{D}((1,3,4,9),(-1,1,-1,1),(-\alpha_{780,1351},-\alpha_{780,1351},0,\alpha_{780,1351}))$	$(-\alpha_{1560,2702},\alpha_{1351,2340})$	53421
9	2	$\mathscr{D}((3,2,11,8),(-1,1,-1,1),(-\alpha_{780,1351},\alpha_{1351,2340},0,-\alpha_{1351,2340}))$	$(-\alpha_{2340,4053},\alpha_{2702,4680})$	53575
10	3	$\mathscr{D}((3,8,11),(-1,-1,-1),(-\alpha_{780,1351},-\alpha_{1351,2340},0))$	$(-\alpha_{2340,4053},\alpha_{2702,4680})$	53707
11	2	$\mathscr{D}((4,2,11,8),(-1,1,-1,1),(\alpha_{7173,12424},\alpha_{3482,6031},0,-\alpha_{3482,6031}))$	$(-\alpha_{6031,1044},\alpha_{6964,12062})$	608839
12	3	$\mathscr{D}((4,8,11),(-1,-1,-1),(\alpha_{7173,12424},-\alpha_{3482,6031},0))$	$(-\alpha_{6031,1044},\alpha_{6964,12062})$	608971
13	2	$\mathscr{D}((6,4,3,10),(-1,1,-1,1),(\alpha_{3691,6393},\alpha_{7173,12424},-\alpha_{12424,21519},-\alpha_{7173,12424}))$	$(-\alpha_{12424,21519},\alpha_{7173.12424})$	610259
14	4	$\mathscr{D}((6,10,11,3),(-1,-1,-1,-1),(\alpha_{3691,6393},-\alpha_{7173,12424},0,-\alpha_{1560,2702}))$	$(-\alpha_{5251,9095},\alpha_{1922,3329})$	348666
15	3	$\mathscr{D}((2,6,11),(-1,-1,-1),(\alpha_{3482,6031},\alpha_{3691,6393},0))$	$(-\alpha_{6031,10446},\alpha_{6964,12062})$	608971
16	2	$\mathscr{D}((11,0,3,6),(-1,1,-1,1),(0,-\alpha_{3691,6393},-\alpha_{1560,2702},\alpha_{3691,6393}))$	$(0,-\alpha_{3691,6393})$	402015
17	4	$\mathscr{D}((0,3,8,11),(-1,-1,-1,-1),(-\alpha_{3691,6393},-\alpha_{1560,2702},-\alpha_{571,989},0))$	$(-\alpha_{989,1713},-\alpha_{989,1713})$	244544
18	2	$\mathscr{D}((1,3,4,9),(-1,1,-1,1),(-\alpha_{1560,2702},-\alpha_{1560,2702},\alpha_{2131,3691},\alpha_{1560,2702}))$	$(-\alpha_{3120,5404},\alpha_{2702,4680})$	80613
19	4	$\mathscr{D}((4,8,9,1),(-1,-1,-1,-1),(0,-\alpha_{571,989},\alpha_{1560,2702},-\alpha_{1560,2702}))$	$(-\alpha_{2131,3691},\alpha_{2131,3691})$	53630
20	2	$\mathscr{D}((2,1,11,7),(-1,1,-1,1),(\alpha_{1351,2340},-\alpha_{1560,2702},0,\alpha_{1560,2702}))$	$(-\alpha_{1560,2702},\alpha_{2702,4680})$	53993
21	3	$\mathscr{D}((2,7,11),(-1,-1,-1),(\alpha_{1351,2340},\alpha_{1560,2702},0))$	$(-\alpha_{2340,4053},\alpha_{2702,4680})$	53707
22	3	$\mathscr{D}((4,11,7),(-1,-1,-1),(0,0,\alpha_{780,1351}))$	$(-\alpha_{780,1351},\alpha_{1351,2340})$	26933

A.2.3 *Level-0 periodic domains, sequence A*

j	#V	$\mathscr{D}(\mathbf{m},\mathbf{s},\mathbf{b})$	$T_j(0)$
0	4	$\mathscr{D}((4,11,8,7),(-1,-1,-1,-1),(-\alpha_{10864,18817},-\alpha_{6602,11435},\alpha_{21157,36645},\alpha_{5251,9095}))$	2499612
1	4	$\mathscr{D}((0,7,4,3),(-1,-1,-1,-1),(\alpha_{571,989},\alpha_{209,362},0,\alpha_{1351,2340}))$	81276
2	6	$\mathscr{D}((7,6,3,2,11,10),(-1,-1,-1,-1,-1,-1),(-\alpha_{571,989},-\alpha_{1351,2340},-\alpha_{4471,7744},\alpha_{571,989},0,-\alpha_{4262,7382}))$	174436
3	4	$\mathscr{D}((0,1,4,9),(-1,-1,-1,-1),(-\alpha_{780,1351},\alpha_{4262,7382},\alpha_{5042,8733},\alpha_{3691,6393}))$	644484
4	3	$\mathscr{D}((0,4,7),(-1,-1,-1),(-\alpha_{780,1351},\alpha_{5042,8733},-\alpha_{4262,7382}))$	1092180
5	3	$\mathscr{D}((1,5,10),(-1,-1,-1),(\alpha_{1351,2340},-\alpha_{2131,3691},-\alpha_{7953,13775}))$	1092180
6	4	$\mathscr{D}((0,3,7,8),(-1,-1,-1,-1),(-\alpha_{780,1351},-\alpha_{780,1351},-\alpha_{4262,7382},-\alpha_{4262,7382}))$	563340
7	4	$\mathscr{D}((0,3,8,11),(-1,-1,-1,-1),(-\alpha_{3691,6393},-\alpha_{1560,2702},-\alpha_{571,989}))$	351804

A.2.4 *Miscellaneous periodic domains, $j > 10$*

j	#V	$\mathscr{D}(\mathbf{m},\mathbf{s},\mathbf{b})$	$T_j(0)$
11	4	$\mathscr{D}((0,1,6,7),(-1,-1,-1,-1),(\alpha_{2,4},\alpha_{2,4},\alpha_{2,3},\alpha_{2,3}))$	30
12	12	$\mathscr{D}((0,\ldots,11),(-1)^{12},(\alpha_{155,269},\alpha_{1480,2564},\alpha_{192,333},\alpha_{1098,1902},\alpha_{817,1415},\alpha_{398,689},\alpha_{1405,2433},\alpha_{80,138},\alpha_{1368,2369},\alpha_{462,800},\alpha_{743,1287},\alpha_{1162,2013}))$	13001
13	12	$\mathscr{D}((0,\ldots,11),(-1)^{12},(\alpha_{2,4},-\alpha_{80,138},-\alpha_{17,29},-\alpha_{44,76},-\alpha_{60,104},-\alpha_{5,9},-\alpha_{84,146},-\alpha_{2,4},-\alpha_{65,113},-\alpha_{38,66},-\alpha_{22,38},-\alpha_{77,133}))$	355
14	12	$\mathscr{D}((0,\ldots,11),(-1)^{12},(\alpha_{2,4},\alpha_{114,198},\alpha_{9,16},\alpha_{79,137},\alpha_{63,109},\alpha_{21,36},\alpha_{110,190},-\alpha_{2,4},\alpha_{103,178},\alpha_{33,57},\alpha_{49,85},\alpha_{91,158}))$	473
15	12	$\mathscr{D}((0,\ldots,11),(-1)^{12},(-\alpha_{4,6},-\alpha_{4,6},-\alpha_{1,1},-\alpha_{6,10},0,-\alpha_{5,9},-\alpha_{2,4},-\alpha_{2,4},-\alpha_{5,9},0,-\alpha_{6,10},-\alpha_{1,1}))$	7
16	8	$\mathscr{D}((1,2,4,5,7,8,10,11),(-1)^{8},(\tau,-\alpha_{3,4},-\alpha_{1,1},-\alpha_{3,5}-\alpha_{3,6},-\tau,-\alpha_{2,4},0))$	9
17	6	$\mathscr{D}((6,7,10,11,2,3),(-1,-1,-1,-1,-1,-1),(-\alpha_{41,72},-\alpha_{250,434},-\alpha_{55,95},-\alpha_{113,195},-\alpha_{266,460},\alpha_{1,2}))$	15524

A.2.5 Level-0 scaling domains, sequence B

j	#V	$\mathscr{D}(\mathbf{m},\mathbf{s},\mathbf{b})$	$T_j(0)$	$\mathbf{d}_j(0)$
0	3	$\mathscr{D}((6,10,3),(-1,-1,-1),(0,\alpha_{571,989},\alpha_{209,362}))$	19459	$(\alpha_{209,362},-\alpha_{362,627})$
1	2	$\mathscr{D}((6,2,11,8),(-1,1,-1,1),(0,-\alpha_{571,989},0,\alpha_{571,989}))$	57089	$(0,-\alpha_{571,989})$
2	2	$\mathscr{D}((2,1,9,7),(-1,1,-1,1),(-\alpha_{1351,2340},\alpha_{780,1351},-\alpha_{780,1351},-\alpha_{780,1351}))$	86787	$(-\alpha_{780,1351},0)$
3	4	$\mathscr{D}((6,7,11,2),(-1,-1,-1,-1),(0,-\alpha_{780,1351},0,-\alpha_{571,989}))$	76580	$(-\alpha_{571,989},\alpha_{209,362})$
4	2	$\mathscr{D}((6,3,1,9),(-1,1,-1,1),(0,\alpha_{780,1351},\alpha_{780,1351},-\alpha_{780,1351}))$	92983	$(\alpha_{780,1351},-\alpha_{1351,2340})$
5	3	$\mathscr{D}((1,6,9),(1,-1,-1),(\alpha,0,-\alpha))$	93115	$(\alpha_{780,1351},-\alpha_{1351,2340})$
6	3	$\mathscr{D}((6,11,3),(1,-1,-1),(0,0,\alpha_{780,1351}))$	93115	$(\alpha_{780,1351},-\alpha_{1351,2340})$
7	2	$\mathscr{D}((0,4,6,10),(-1,1,-1,1),(\alpha_{780,1351},-\alpha_{571,989},0,\alpha_{571,989}))$	58245	$(\alpha_{989,1713},-\alpha_{1142,1978})$
8	4	$\mathscr{D}((3,11,8,5),(-1,-1,-1,-1),(\alpha_{209,362},0,\alpha_{571,989},\alpha_{780,1351}))$	38918	$(\alpha_{571,989},-\alpha_{780,1351})$
9	2	$\mathscr{D}((0,5,8,11),(-1,1,-1,1),(\alpha_{780,1351},\alpha_{780,1351},\alpha_{571,989},-\alpha_{780,1351}))$	58245	$(\alpha_{989,1713},-\alpha_{1142,1978})$
10	3	$\mathscr{D}((4,11,8),(-1,-1,-1),(-\alpha_{571,989},-\alpha_{780,1351},\alpha_{571,989}))$	58377	$(\alpha_{989,1713},-\alpha_{1142,1978})$

A.2.6 Level-0 periodic domains, sequence B

j	#V	$\mathscr{D}(\mathbf{m},\mathbf{s},\mathbf{b})$	$T_j(0)$
0	4	$\mathscr{D}((3,10,7,6),(-1,-1,-1,-1),(\alpha_{209,362},\alpha_{571,989},\alpha_{1351,2340},0))$	233508
1	6	$\mathscr{D}((6,7,10,11,2,3),(-1,-1,-1,-1,-1,-1),(0,-\alpha_{780,1351},-\alpha_{780,1351},0,-\alpha_{571,989},-\alpha_{571,989}))$	76580

A.2.7 Incidence matrices

Sequence A

$$
\mathbf{A}=\begin{pmatrix}
0&0&0&0&0&0&0&0&0&0&0&4&4&4&2&5&2&1&0&0&0&0&0\\
0&0&0&0&0&0&0&0&0&0&0&1&0&0&0&0&0&0&0&0&0&0&0\\
0&0&0&0&0&0&0&0&0&0&0&24&24&24&12&25&14&7&0&0&0&0&0\\
0&0&0&0&0&0&0&0&0&0&0&0&0&0&0&0&0&0&0&0&0&0&0\\
0&0&0&0&0&0&0&0&0&0&0&4&5&4&2&4&2&1&0&0&0&0&0\\
0&0&0&0&0&0&0&0&0&0&0&0&0&1&0&0&1&0&0&0&0&0&0\\
0&1&0&1&0&0&0&0&0&0&0&0&0&0&0&0&0&0&0&0&0&0&0\\
0&1&1&1&2&1&1&1&1&1&1&1&1&1&1&1&1&1&1&1&1&1&1\\
0&0&0&0&0&1&0&0&1&0&0&0&0&0&0&0&0&0&0&0&0&0&0\\
0&0&0&0&0&0&1&0&0&1&0&0&0&0&0&0&0&0&0&0&0&0&0\\
0&0&0&0&0&0&0&1&1&1&2&1&1&1&1&1&1&1&1&1&1&1&0\\
0&0&0&0&0&0&0&0&0&0&0&0&0&0&0&0&0&0&0&0&0&0&0\\
0&0&0&0&0&0&0&0&0&0&0&0&0&0&0&0&0&0&0&0&0&0&0\\
0&0&0&0&0&0&0&0&0&0&0&0&0&0&0&0&0&0&0&0&0&0&0\\
0&0&0&0&0&0&0&0&0&0&0&0&0&0&0&0&0&0&0&0&0&0&0\\
0&0&0&0&0&0&0&0&0&0&0&0&0&0&0&0&0&0&0&0&0&0&0\\
0&0&0&0&0&0&0&0&0&0&0&0&0&0&0&0&0&0&0&0&0&0&0\\
0&0&0&0&0&0&0&0&0&0&0&0&0&0&0&0&0&0&0&0&0&0&0\\
0&0&0&0&0&0&0&0&0&0&0&0&0&0&0&0&0&0&0&0&0&0&0\\
0&0&0&0&0&0&0&0&0&0&0&34&34&34&19&34&22&13&2&1&0&0&0\\
0&0&0&0&0&0&0&0&0&0&0&0&0&0&0&0&0&0&1&0&1&0&0\\
0&0&0&0&0&0&0&0&1&1&1&1&1&1&1&1&1&1&1&1&1&2&1\\
1&1&1&1&1&1&1&1&1&1&1&1&1&1&1&1&1&1&1&1&1&1&1
\end{pmatrix} \qquad \text{(A.1)}
$$

Sequence B

$$
\mathbf{A}=\begin{pmatrix}
0&0&0&0&0&0&1&0&0&0&0\\
0&0&1&0&0&0&0&0&0&0&0\\
0&1&0&0&0&0&0&0&0&0&0\\
0&0&2&2&2&2&2&0&0&0&0\\
0&0&0&0&0&0&0&1&0&1&0\\
0&1&1&1&1&1&1&1&1&1&2\\
1&1&1&1&1&1&1&1&1&1&1\\
0&0&0&0&1&0&0&0&0&0&0\\
0&0&2&1&2&2&3&0&0&0&0\\
0&0&0&0&0&0&0&0&0&0&0\\
0&0&0&0&0&1&0&0&0&0&0
\end{pmatrix} \qquad \text{(A.2)}
$$

A.3 Model III Data Tables, from Kouptsov et al. (2002)

A.3.1 Generating domain

Below is a table of generating domains and subdomains, and the base domain for the scaling sequence. The coordinates for the base domain are relative to the scaling limit point $\mathbf{x}_\infty$.

Domain	#V	$\mathscr{D}(\mathbf{m},\mathbf{s},\mathbf{b})$	$\mathbf{d}_j(\Omega)$
Ω	4	$\mathscr{D}((0,1,0,1),(1,1,-1,-1),(-\tau,-\tau,\tau,\tau))$	——
Ω_0	3	$\mathscr{D}((1,4,6),(-1,-1,-1),(\tau,\tau,-\tau))$	$(0,-2\tau)$
Ω_1	6	$\mathscr{D}((0,1,2,4,5,6),(-1,-1,-1,-1,-1,-1),(\tau,\tau,\tau,\tau,\tau,\tau))$	$(0,0)$
Ω_2	3	$\mathscr{D}((0,5,2),(-1,-1,-1),(\tau,\tau,-\tau))$	$(0,2\tau)$
Ω_3	2	$\mathscr{D}((2,0,6,5),(1,-1,1,-1),(-\tau,\tau,\tau,\tau))$	$(0,0)$
Ω_4	2	$\mathscr{D}((1,0,5,4),(1,-1,1,-1),(-\tau,\tau,\tau,-1-\tau))$	$(0,2\tau)$
Ω_5	1	$\mathscr{D}((1,2,5,6),(1,1,1,1),(-\tau,-\tau,\tau,\tau))$	$(0,0)$
Ω_6	2	$\mathscr{D}((1,0,5,4),(1,-1,1,-1),(-\tau,1+\tau,\tau,\tau))$	$(0,0)$
Ω_7	1	$\mathscr{D}((0,1,4,5),(1,1,1,1),(-\tau,-\tau,\tau,\tau))$	$(0,0)$
Ω_8	2	$\mathscr{D}((0,1,4,5),(1,-1,1,-1),(-\tau,0,\tau,\tau))$	$(0,0)$
Ω_9	1	$\mathscr{D}((0,1,4,5),(1,1,1,1),(-\tau,0,\tau,0))$	$(0,-2\tau)$
Ω_{10}	2	$\mathscr{D}((0,1,4,5),(1,-1,1,-1),(-\tau,\tau,\tau,0))$	$(0,-2\tau)$
Ω_{11}	2	$\mathscr{D}((2,1,6,5),(1,-1,1,-1),(\tau,\tau,-\tau,0))$	$(0,-2\tau)$
$D(-1)$	3	$\mathscr{D}((1,4,6),(-1,-1,-1),(0,0,-1))$	——

A.3.2 Pre scaling level $L=-1$

Return map data for the pre-scaling domain $D(-1)$. Coordinates are relative to the scaling limit point $\mathbf{x}_\infty = (-1-\tau,\tau)$.

j	$\mathscr{D}(\mathbf{m},\mathbf{s},\mathbf{b})$	$\mathbf{d}_j(-1)$	$T_j(-1)$
0	$\mathscr{D}((3,4,7,0),(1,-1,1,-1),(-1,-1,1,3+3\tau))$	$(-3-6\tau,4+6\tau)$	13
1	$\mathscr{D}((0,1,4,5),(1,1,1,1),(3+3\tau,-4-6\tau,-3-3\tau,4+6\tau))$	$(7+10\tau,-5-6\tau)$	22
2	$\mathscr{D}((3,5,7,1),(1,-1,1,-1),(-1,4+6\tau,1,0))$	$(4+4\tau,-4-6\tau)$	12
3	$\mathscr{D}((0,1,3,4,6),(-1,-1,-1,-1,-1),(-1-2\tau,0,-1,-1,0))$	$(1,1)$	2
4	$\mathscr{D}((0,2,4,6),(1,1,1,1),(-1-2\tau,-3-4\tau,1+2\tau,3+4\tau))$	$(1,0)$	13
5	$\mathscr{D}((0,1,4,7),(1,-1,1,-1),(-1-2\tau,0,1+2\tau,-1-2\tau))$	$(-1-2\tau,1+2\tau)$	6
6	$\mathscr{D}((0,2,4,6),(1,-1,1,-1),(-1-2\tau,-3-4\tau,1+2\tau,0))$	$(5+6\tau,-4-6\tau)$	13
7	$\mathscr{D}((1,6,4),(-1,-1,-1),(0,0,1+2\tau))$	$(2+2\tau,-1-2\tau)$	3
8	$\mathscr{D}((4,1,7),(-1,-1,-1),(-1,0,1))$	$(2,1)$	3

A.3.3 Domains $\mathbf{D}_j(L)$ for even L

j	$\mathscr{D}(\mathbf{m},\mathbf{s},\mathbf{b})$	$\mathbf{d}_j(L)$	$T_j(L)$
0	$\mathscr{D}((4,1,7,6),(-1,-1,-1,-1),(-\alpha^L_{1,-2},0,\alpha^L_{1,0},0))$	$(\alpha^L_{2,-2},\alpha^L_{1,-2})$	3×9^L
1	$\mathscr{D}((3,1,7,6),(1,-1,1,-1),(-\alpha^L_{1,0},\alpha^L_{2,2},\alpha^L_{1,0},0))$	$(\alpha^L_{2,2},-\alpha^L_{2,2})$	$36\times 9^L-4\times 5^L$
2	$\mathscr{D}((1,2,5,6),(1,1,1,1),(\alpha^L_{3,4},-\alpha^L_{2,2},-\alpha^L_{3,4},\alpha^L_{2,2}))$	$(-\alpha^L_{0,4},\alpha^L_{2,2})$	$\frac{177}{2}\times 9^L-11\times 5^L-\frac{3}{2}$
3	$\mathscr{D}((3,1,7,6),(1,-1,1,-1),(-\alpha^L_{1,0},0,\alpha^L_{1,0},\alpha^L_{2,2}))$	$(-\alpha^L_{0,4},\alpha^L_{2,2})$	$66\times 9^L-6\times 5^L$
4	$\mathscr{D}((0,1,4,5),(1,1,1,1),(-\alpha^L_{1,3},\alpha^L_{2,2},\alpha^L_{1,3},-\alpha^L_{2,2}))$	$(-\alpha^L_{1,4},\alpha^L_{1,2})$	$96\times 9^L-10\times 5^L$
5	$\mathscr{D}((0,1,4,6),(1,-1,1,-1),(-\alpha^L_{1,2},0,\alpha^L_{1,2},0))$	$(\alpha^L_{1,-2},\alpha^L_{2,2})$	$75\times 9^L-6\times 5^L$
6	$\mathscr{D}((0,1,3,6),(-1,-1,-1,-1),(-\alpha^L_{1,2},0,-\alpha^L_{1,0},0))$	$(-\alpha^L_{0,2},\alpha^L_{1,2})$	15×9^L
7	$\mathscr{D}((3,5,7,2),(1,-1,1,-1),(-\alpha^L_{1,0},-\alpha^L_{2,2},\alpha^L_{1,0},-\alpha^L_{2,2}))$	$(-\alpha^L_{1,4},\alpha^L_{2,2})$	$5\times 9^L-4\times 5^L$
8	$\mathscr{D}((4,1,6),(-1,-1,-1),(\alpha^L_{1,2},0,0))$	$(\alpha^L_{2,0},0)$	24×9^L

A.3.4 Domains $\mathbf{D}_j(L)$ for odd L

j	$\mathscr{D}(\mathbf{m},\mathbf{s},\mathbf{b})$	$\mathbf{d}_j(-1)$	$T_j(L)$
0	$\mathscr{D}((4,1,7,6),(-1,-1,-1,-1),(-\alpha^L_{1,-2},0,\alpha^L_{1,0},0))$	$(\alpha^L_{2,-2},\alpha^L_{1,-2})$	3×9^L
1	$\mathscr{D}((3,1,7,4),(1,-1,1,-1),(-\alpha^L_{1,0},0,\alpha^L_{1,0},-\alpha^L_{1,0}))$	$(-\alpha^L_{0,2},\alpha^L_{2,2})$	$36\times 9^L-4\times 5^L$
2	$\mathscr{D}((0,2,4,6),(1,1,1,1),(\alpha^L_{1,0},\alpha^L_{1,2},-\alpha^L_{1,0},-\alpha^L_{1,2}))$	$(-\alpha^L_{0,4},\alpha^L_{2,2})$	$\frac{177}{2}\times 9^L-11\times 5^L-\frac{3}{2}$
3	$\mathscr{D}((3,1,7,0),(1,-1,1,-1),(-\alpha^L_{1,0},-\alpha^L_{2,2},\alpha^L_{1,0},\alpha^L_{1,0}))$	$(-\alpha^L_{0,4},\alpha^L_{2,2})$	$66\times 9^L-6\times 5^L$
4	$\mathscr{D}((0,1,4,5),(1,1,1,1),(-\alpha^L_{1,3},\alpha^L_{2,2},\alpha^L_{1,3},-\alpha^L_{2,2}))$	$(-\alpha^L_{1,4},\alpha^L_{1,2})$	$96\times 9^L-10\times 5^L$
5	$\mathscr{D}((3,1,7,6),(1,-1,1,-1),(-\alpha^L_{1,0},\alpha^L_{2,2},\alpha^L_{1,0},0))$	$(\alpha^L_{3,0},0)$	$75\times 9^L-6\times 5^L$
6	$\mathscr{D}((0,1,3,6),(-1,-1,-1,-1),(-\alpha^L_{1,2},0,-\alpha^L_{1,0},0))$	$(-\alpha^L_{0,2},\alpha^L_{1,2})$	15×9^L
7	$\mathscr{D}((0,6,4,1),(1,-1,1,-1),(-\alpha^L_{1,2},0,\alpha^L_{1,2},0))$	$(-\alpha^L_{1,4},\alpha^L_{2,2})$	$5\times 9^L-4\times 5^L$
8	$\mathscr{D}((4,1,6),(-1,-1,-1),(\alpha^L_{1,2},0,0))$	$(\alpha^L_{2,0},0)$	24×9^L

A.3.5 Domains $\boldsymbol{\Pi}_j(L)$ for all L

L	j	$\mathscr{D}(\mathbf{m},\mathbf{s},\mathbf{b})$	$\tilde{T}_j(L)$
-1	0	$\mathscr{D}((0,\dots,7),(-1)^8,(-1-2\tau,0,-2-2\tau,-1,-1,-2-2\tau,0,-1-2\tau))$	2
	1	$\mathscr{D}((0,1,4,7),(1,-1,1,-1),(-1-2\tau,0,1+2\tau,-1-2\tau))$	8
≥ 0	0	$\mathscr{D}((0,\dots,7),(-1)^8,(\alpha^L_{1,0},0,\alpha^L_{0,2},-\alpha^L_{1,-2},-\alpha^L_{1,-2},\alpha^L_{0,2},0,\alpha^L_{1,0}))$	$3\cdot 9^L$
	1	$\mathscr{D}((0,\dots,7),(-1)^8,(-\alpha^L_{1,2},0,-\alpha^L_{2,2},-\alpha^L_{1,0},-\alpha^L_{1,0},-\alpha^L_{2,2},0,-\alpha^L_{1,2}))$	$18\cdot 9^L$
	2	$\mathscr{D}((0,\dots,7),(-1)^8,(\alpha^L_{5,6},0,\alpha^L_{2,4},\alpha^L_{3,6},-\alpha^L_{1,0},\alpha^L_{4,6},\alpha^L_{2,2},\alpha^L_{1,0}))$	$27\cdot 9^L$
even	3	$\mathscr{D}((3,0,7,4),(1,-1,1,-1),(-\alpha^L_{1,0},-\alpha^L_{2,4},\alpha^L_{1,0},-\alpha^L_{1,0}))$	$84\cdot 9^L-4\cdot 5^L$
odd	3	$\mathscr{D}((0,1,4,5),(1,-1,1,-1),(-\alpha^L_{1,2},0,\alpha^L_{1,2},-\alpha^L_{3,4}))$	$84\cdot 9^L-4\cdot 5^L$

A.3.6 Tiling data

D	$\mathscr{A}(D)$	$\mathscr{W}(D)$	$\nu(D)$
$D_0(-1)$	0	$-2-3\tau$	13
$D_1(-1)$	0	0	22
$D_2(-1)$	0	$3+4\tau$	12
$D_3(-1)$	$2+\frac{5}{2}\tau$	$-\frac{3}{2}-\tau$	2
$D_4(-1)$	0	0	13
$D_5(-1)$	0	$3+4\tau$	6
$D_6(-1)$	0	$-2-3\tau$	13
$D_7(-1)$	$-1-\frac{3}{2}\tau$	$-\frac{1}{2}$	3
$D_8(-1)$	$-1-\frac{3}{2}\tau$	$-\frac{1}{2}$	3
$D_0(0)$	$5+7\tau$	$1+2\tau$	1
$D_1(0)$	0	$5+7\tau$	5
$D_2(0)$	0	0	10
$D_3(0)$	0	$-12-17\tau$	11
$D_4(0)$	0	0	14
$D_5(0)$	0	$5+7\tau$	15
$D_6(0)$	$29+41\tau$	$-3-4\tau$	6
$D_7(0)$	0	$5+7\tau$	8
$D_8(0)$	$-35-\frac{99}{2}\tau$	$-\frac{3}{2}-2\tau$	9
$D_0(1)$	$119+239\tau$	$7+10\tau$	2
$D_1(1)$	0	$29+41\tau$	20
$D_2(1)$	0	0	52
$D_3(1)$	0	$-70-99\tau$	54
$D_4(1)$	0	0	70
$D_5(1)$	0	$29+41\tau$	68

D	$\mathscr{A}(D)$	$\mathscr{W}(D)$	$\nu(D)$
$D_6(1)$	$985+1393\tau$	$-17-24\tau$	18
$D_7(1)$	0	$29+41\tau$	34
$D_8(1)$	$-1189-\frac{3363}{2}\tau$	$-\frac{17}{2}-12\tau$	33
$D_0(2)$	$5741+8119\tau$	$41+58\tau$	2
$D_1(2)$	0	$169+239\tau$	20
$D_2(2)$	0	0	52
$D_3(2)$	0	$-408-577\tau$	54
$D_4(2)$	0	0	70
$D_5(2)$	0	$169+239\tau$	68
$D_6(2)$	$33461+47321\tau$	$-99-140\tau$	18
$D_7(2)$	0	$169+239\tau$	34
$D_8(2)$	$40391+\frac{114243}{2}\tau$	$-\frac{99}{2}-70\tau$	33
$\Pi_0(\Omega)$	$4+4\tau$	$4+8\tau$	1
$\Pi_0(-1)$	$20+28\tau$	$-12-16\tau$	1
$\Pi_1(-1)$	0	$3+4\tau$	2
$\Pi_2(-1)$	$116+164\tau$	$28+40\tau$	1
$\Pi_0(0)$	$116+164\tau$	$28+40\tau$	1
$\Pi_1(0)$	$676+956\tau$	$-68-96\tau$	2
$\Pi_2(0)$	$3940+5572\tau$	$164+232\tau$	5
$\Pi_3(0)$	0	$17+24\tau$	6
$\Pi_0(1)$	$3940+5572\tau$	$164+232\tau$	1
$\Pi_1(1)$	$22964+32476\tau$	$-396-560\tau$	2
$\Pi_2(1)$	$133844+189284\tau$	$956+1352\tau$	5
$\Pi_3(1)$	0	$99+140\tau$	6

A.3.7 Section of the discontinuity set

Generating domains

$$\mathscr{B} = \Omega_{7\smile 8\smile 9\smile 10}$$
$$\Omega_7 = K^7 D_4^{(-1)}$$
$$\Omega_8 = K^{11}D_2^{(-1)} \smile K^{21}D_1^{(-1)} \smile K^{12}D_6^{(-1)} \smile K^8 D_{4\smile 6}^{(-1)} \smile K^{17}D_1^{(-1)} \smile K^7 D_{2\smile 1\smile 0}^{(-1)}$$
$$\Omega_9 = K^{11}D_1^{(-1)}$$
$$\Omega_{10} = K^3 D_{0\smile 1\smile 2}^{(-1)} \smile K^1 3D_1^{(-1)} \smile K^4 D_{6\smile 4\smile 5}^{(-1)} \smile K^{11}D_{4\smile 0}^{(-1)}$$

Pre-scaling domains

$$D_0^{(-1)} = \rho_{-1}^4 D_{7\smile 2\smile 3}^{(0)} \smile \rho_{-1}^7 D_5^{(0)}$$
$$D_1^{(-1)} = \rho_{-1}^4 D_4^{(0)}$$
$$D_2^{(-1)} = \rho_{-1}^5 D_5^{(0)} \smile \rho_{-1}^7 D_2^{(0)} \smile \rho_{-1}^{14} D_5^{(0)} \smile \rho_{-1}^{13} D_4^{(0)} \smile \rho_{-1}^{10} D_3^{(0)} \smile \rho_{-1}^9 D_2^{(0)} \smile \rho_{-1}^4 D_1^{(0)}$$
$$D_4^{(-1)} = \rho_{-1}^8 D_2^{(0)}$$
$$D_5^{(-1)} = \Pi_1^{(-1)}$$
$$D_6^{(-1)} = \rho_{-1} D_{1\smile 4\smile 7\smile 2\smile 3}^{(0)}$$

Scaling sequence, *L* even

$$\begin{aligned}
D_1^{(L)} &= \rho_L^{52} D_5^{(L+1)} \smile \rho_L^{20} D_2^{(L+1)} \smile \rho_L^{7} D_{5\smile4\smile3\smile2\smile1}^{(L+1)} \\
D_2^{(L)} &= \rho_L^{11} D_2^{(L+1)} \\
D_3^{(L)} &= \rho_L^{16} D_7^{(L+1)} \smile \rho_L^{34} D_2^{(L+1)} \smile \rho_L^{36} D_{3\smile4\smile5}^{(L+1)} \\
D_4^{(L)} &= \rho_L^{52} D_4^{(L+1)} \\
D_5^{(L)} &= \rho_L^{13} D_1^{(L+1)} \smile \rho_L^{63} D_4^{(L+1)} \smile \rho_L^{27} D_7^{(L+1)} \smile \rho_L^{45} D_2^{(L+1)} \smile \rho_L^{47} D_{3\smile4\smile5}^{(L+1)} \\
&\quad \smile \rho_L^{15} D_2^{(L+1)} \smile \rho_L^{2} D_{5\smile4\smile3\smile2\smile1}^{(L+1)} \\
D_7^{(L)} &= \Pi_3(L) \smile \rho_L^{13} D_2^{(L+1)} \smile \rho_L^{18} D_1^{(L+1)} \smile \rho_L^{68} D_4^{(L+1)} \smile \rho_L^{32} D_7^{(L+1)} \\
&\quad \smile \rho_L^{50} D_2^{(L+1)} \smile \rho_L^{52} D_3^{(L+1)}
\end{aligned}$$

Scaling sequence, *L* odd

$$\begin{aligned}
D_1^{(L)} &= \rho_L^{16} D_5^{(L+1)} \smile \rho_L^{32} D_2^{(L+1)} \smile \rho_L^{61} D_5^{(L+1)} \smile \rho_L^{63} D_4^{(L+1)} \smile \rho_L^{47} D_3^{(L+1)} \\
&\quad \smile \rho_L^{45} D_2^{(L+1)} \smile \rho_L^{13} D_{1\smile4\smile7\smile2\smile3}^{(L+1)} \smile \rho_L^{29} D_4^{(L+1)} \smile \rho_L^{27} D_5^{(L+1)} \\
D_2^{(L)} &= \rho_L^{41} D_2^{(L+1)} \\
D_3^{(L)} &= \rho_L^{32} D_5^{(L+1)} \smile \rho_L^{34} D_4^{(L+1)} \smile \rho_L^{18} D_{3\smile2\smile7}^{(L+1)} \\
D_4^{(L)} &= \rho_L^{18} D_4^{(L+1)} \\
D_5^{(L)} &= \rho_L^{18} D_1^{(L+1)} \smile \rho_L^{50} D_2^{(L+1)} \smile \rho_L^{52} D_3^{(L+1)} \smile \rho_L^{68} D_4^{(L+1)} \smile \rho_L^{66} D_5^{(L+1)} \\
&\quad \smile \rho_L^{37} D_2^{(L+1)} \smile \rho_L^{21} D_5^{(L+1)} \smile \rho_L^{23} D_4^{(L+1)} \smile \rho_L^{7} D_{3\smile2\smile7\smile4\smile1}^{(L+1)} \\
D_7^{(L)} &= \Pi_3(L) \smile \rho_L^{39} D_2^{(L+1)} \smile \rho_L D_{1\smile4\smile7\smile2\smile3}^{(L+1)}
\end{aligned}$$

A.4 Cubic Model Data Tables, from Lowenstein et al. (2004)

Table A.1 Data for dressed subdomains. Each $\Delta(i)$ is generated from its equivalent catalogue member $\Delta^{[h(i)]}$ by application of the transformation $R^{\kappa} \circ C^{-\nu} \circ T_{-\mathbf{d}} \circ S_{\omega^{-1}}$. Algebraic numbers $a + b\lambda + c\lambda^2$ are represented here as $z[a,b,c]$.

$\Delta(i)$	κ	ν	$\mathbf{d}$	ω
$\Delta(0)$	1	8	$(z[1,-1,0], z[-1,0,0])$	$z[-3,5,6]$
$\Delta(1)$	0	13	$(z[2,4,-3], z[2,-3,1])$	$z[-5,6,8]$
$\Delta(2)$	0	8	$(0, z[5,-3,0])$	$z[-14,20,25]$
$\Delta(3)$	1	13	$(z[-9,-2,4], z[-4,-1,2])$	$z[-17,25,31]$
$\Delta(4)$	1	3	$(z[-1,-1,1], z[-8,1,2])$	$z[-11,16,20]$
$\Delta(5)$	0	4	$(z[16,4,-7], z[8,3,-4])$	$z[-14,20,25]$
$\Delta(6)$	0	3	$(z[2,1,-1], z[2,1,-1])$	$z[-2,3,4]$
$\Delta(7)$	0	9	$(z[-11,-3,5], z[-5,-1,2])$	$z[-2,3,4]$
$\Delta(8)$	1	4	$(z[8,3,-4], z[-3,4,5])$	$z[-2,3,4]$
$\Delta(9)$	0	0	$(0,0)$	$z[0,1,1]$
$\Delta(10)$	1	7	$(z[1,1,-1], z[1,-1,0])$	$z[-2,1,2]$
$\Delta(11)$	0	0	$(z[-2,-1,1], 0)$	$z[0,1,1]$
$\Delta(12)$	0	0	$(0,0)$	$z[-2,3,4]$
$\Delta(13)$	1	7	$(z[6,2,-3], z[3,0,-1])$	$z[-3,4,5]$
$\Delta(14)$	0	0	$(z[-5,-1,2], 0)$	$z[0,1,1]$

Table A.2 Data for partition of prototype domain $D^{[n]}$. The return map for the atom $D_j^{[n]}$ is parametrized by ν and **d**.

Atom	**m**	**b**	ν	**d**
$D_0^{[0]}$	$(9,11,1,3,7)$	$(1,\lambda,0,-1,0)$	2	$(1,1)$
$D_1^{[0]}$	$(1,11,7)$	$(0,1,0)$	0	$(1,0)$
$D_2^{[0]}$	$(4,1,10)$	$(0,1,0)$	3	$(1,1)$
$D_3^{[0]}$	$(9,4,1)$	$(1,-\lambda,0)$	3	$(2,\lambda)$
$D_0^{[1]}$	$(3,1,11,7)$	$(-2-\lambda+\lambda^2,0,1,0)$	10	$(-2-\lambda+\lambda^2,-1)$
$D_1^{[1]}$	$(11,1,3,7)$	$(2+\lambda-\lambda^2,0,-5-\lambda+2\lambda^2)$	10	$(-5-\lambda+2\lambda^2,-2-\lambda+\lambda^2)$
$D_2^{[1]}$	$(10,13,1,5,7)$	$(5+\lambda-2\lambda^2,-3+\lambda^2,0,-2+\lambda,0)$	12	$(-2+\lambda,3-\lambda^2)$
$D_3^{[1]}$	$(4,1,12,7)$	$(-3-2\lambda+2\lambda^2,0,2-\lambda,0)$	0	$(0,-2+\lambda)$
$D_4^{[1]}$	$(10,1,4,7)$	$(2+\lambda-\lambda^2,0,-2-\lambda+\lambda^2,-2+\lambda)$	0	$(-2+\lambda,0)$
$D_5^{[1]}$	$(10,6,4,0)$	$(2+\lambda-\lambda^2,-2\lambda+\lambda^2,-2-\lambda+\lambda^2,2-\lambda)$	2	$(-1-\lambda+\lambda^2,2\lambda-\lambda^2)$
$D_6^{[1]}$	$(4,12,10)$	$(-2-\lambda+\lambda^2,7-2\lambda^2,2+\lambda-\lambda^2)$	0	$(3+2\lambda-2\lambda^2,3+2\lambda-2\lambda^2)$
$D_7^{[1]}$	$(10,6,1)$	$(5+\lambda-2\lambda^2,3-\lambda^2)$	8	$(0,-5-\lambda+2\lambda^2)$
$D_8^{[1]}$	$(3,1,11,7)$	$(-8-\lambda+3\lambda^2,0,3+2\lambda-2\lambda^2)$	10	$(-8-\lambda+3\lambda^2,-3-2\lambda+2\lambda^2)$
$D_9^{[1]}$	$(13,10,5)$	$(2\lambda-\lambda^2,2+\lambda-\lambda^2,-7+2\lambda^2)$	12	$(-7+2\lambda^2,-2\lambda+\lambda^2)$
$D_{10}^{[1]}$	$(1,10,7)$	$(0,8+\lambda-3\lambda^2,0)$	12	$(3+2\lambda-2\lambda^2,6+2\lambda-3\lambda^2)$
$D_0^{[2]}$	$(9,11,13,1,5,7)$	$(1,-\lambda+\lambda^2,1,0,1-\lambda,0)$	4	$(-\lambda+\lambda^2,-1+\lambda)$
$D_1^{[2]}$	$(4,9,0)$	$(\lambda-\lambda^2,1,1-2\lambda+\lambda^2)$	2	$(1,-1+3\lambda-\lambda^2)$
$D_2^{[2]}$	$(1,9,7)$	$(0,1,-1+2\lambda-\lambda^2)$	0	$(-1+2\lambda-\lambda^2,0)$
$D_3^{[2]}$	$(9,12,1,4,7)$	$(1-2\lambda+\lambda^2,-1+\lambda,0,-1+2\lambda-\lambda^2,0)$	0	$(2\lambda-\lambda^2,0)$
$D_4^{[2]}$	$(3,1,11,7)$	$(2\lambda+\lambda^2,0,1-2\lambda+\lambda^2,0)$	2	$(1,2\lambda-\lambda^2)$
$D_5^{[2]}$	$(1,9,7)$	$(0,2-\lambda,0)$	0	$(-1+\lambda,0)$
$D_6^{[2]}$	$(2,12,7)$	$(-1+2\lambda-\lambda^2,-1+\lambda,0)$	12	$(1+\lambda-\lambda^2,1-\lambda)$
$D_7^{[2]}$	$(10,2,5,7)$	$(2\lambda-\lambda^2,-2+\lambda,3-\lambda^2,0)$	10	$(-1-\lambda+\lambda^2,1+\lambda-\lambda^2)$
$D_8^{[2]}$	$(1,11,7)$	$(0,-\lambda+\lambda^2,-2-\lambda+\lambda^2)$	6	$(-\lambda+\lambda^2,2+\lambda-\lambda^2)$
$D_9^{[2]}$	$(11,6,1)$	$(-\lambda+\lambda^2,-1,1+3\lambda-2\lambda^2)$	10	$(-2-3\lambda+2\lambda^2,1+4\lambda-3\lambda^2)$
$D_{10}^{[2]}$	$(8,6,0)$	$(-1-3\lambda+2\lambda^2,-1,2+\lambda-\lambda^2)$	12	$(3-4\lambda+\lambda^2,3\lambda-2\lambda^2)$
$D_{11}^{[2]}$	$(2,12,7)$	$(-2+\lambda,-3+\lambda^2,0)$	12	$(2+\lambda-\lambda^2,3-\lambda^2)$
$D_0^{[3]}$	$(9,1,3,7)$	$(1,0,1-\lambda,0)$	8	$(0,-1)$
$D_1^{[3]}$	$(10,13,1,4,7)$	$(-1+\lambda,2+\lambda-\lambda^2,0,1+\lambda-\lambda^2,0)$	12	$(0,-2-\lambda+\lambda^2)$
$D_2^{[3]}$	$(10,6,1)$	$(-1+\lambda,-2-\lambda+\lambda^2,0)$	0	$(-3+\lambda^2,0)$
$D_3^{[3]}$	$(11,1,5,7)$	$(-1-\lambda+\lambda^2,0,3-\lambda^2,0)$	10	$(-1-\lambda+\lambda^2,1+\lambda-\lambda^2)$
$D_4^{[3]}$	$(9,12,1,4,7)$	$(2-\lambda,-3+\lambda^2,0,-2+\lambda,0)$	0	$(-1+\lambda,0)$
$D_5^{[3]}$	$(3,1,11,7)$	$(-3-2\lambda+2\lambda^2,0,2-\lambda,0)$	2	$(1,3+2\lambda-2\lambda^2)$
$D_6^{[3]}$	$(1,9,7)$	$(0,8+\lambda-3\lambda^2,0)$	0	$(-7-\lambda+3\lambda^2,0)$
$D_7^{[3]}$	$(2,12,7)$	$(-2+\lambda,-3+\lambda^2,0)$	12	$(2+\lambda-\lambda^2,3-\lambda^2)$
$D_8^{[3]}$	$(10,2,5,7)$	$(3+2\lambda-2\lambda^2,-8-\lambda+3\lambda^2,14+3\lambda-6\lambda^2,0)$	10	$(-10-3\lambda+5\lambda^2,6+2\lambda-3\lambda^2)$
$D_9^{[3]}$	$(2,12,7)$	$(-8-\lambda+3\lambda^2,-14-3\lambda+6\lambda^2,0)$	12	$(7+2\lambda-3\lambda^2,14+3\lambda-6\lambda^2)$

Table A.3 Generating cells $\Pi_j^{[n]}$ constructed from the prototype Π of (2.11) by a transformation $C^{-\nu} \circ T_{-\mathbf{d}} \circ S_{\omega^{-1}}$. The cell is mapped into itself after $\tilde{T}_j^{[n]}$ iterations of $\rho^{[n]}$. Algebraic numbers $a + b\lambda + c\lambda^2$ are represented here as $z[a,b,c]$.

$\Pi_j^{[n]}$	ν	$\mathbf{d}$	ω	$\tilde{T}_j^{[n]}$
$\Pi_0^{[0]}$	1	$(z[5,1,-2],z[1,0,0])$	$z[-2,3,4]$	3
$\Pi_1^{[0]}$	0	$(z[13,5,-7],z[8,1,-3])$	$z[-20,29,36]$	18
$\Pi_2^{[0]}$	0	$(z[-74,-17,32],z[8,1,-3])$	$z[-409,591,737]$	2160
$\Pi_3^{[0]}$	0	$(z[-9,1,2],z[-1,-3,2])$	$z[-45,65,81]$	51
$\Pi_4^{[0]}$	1	$(z[-75,-25,37],z[-28,-2,10])$	$z[-409,591,737]$	264
$\Pi_5^{[0]}$	1	$(z[-380,-99,172],z[-28,-2,10])$	$z[-1328,1919,2393]$	861
$\Pi_6^{[0]}$	1	$(z[-14,-3,6],z[-6,2,1])$	$z[-45,65,81]$	117
$\Pi_7^{[0]}$	1	$(z[60,10,-24],z[33,11,-16])$	$z[-182,263,328]$	111
$\Pi_8^{[0]}$	1	$(z[-5,-8,6],z[-6,2,1])$	$z[-25,36,45]$	33
$\Pi_9^{[0]}$	1	$(z[-49,-16,24],z[-15,-2,6])$	$z[-81,117,146]$	63
$\Pi_{10}^{[0]}$	1	$(z[1,5,-3],z[3,-3,1])$	$z[-6,9,11]$	2
$\Pi_{11}^{[0]}$	1	$(z[-11,-19,14],z[4,-9,4])$	$z[-126,182,227]$	396
$\Pi_{12}^{[0]}$	0	$(z[9,-7,1],z[2,8,-5])$	$z[-56,81,101]$	40
$\Pi_{13}^{[0]}$	0	$(z[-52,-20,27],z[2,8,-5])$	$z[-409,591,737]$	272
$\Pi_{14}^{[0]}$	0	$(z[-4,-16,10],z[-7,13,-5])$	$z[-126,182,227]$	76
$\Pi_{15}^{[0]}$	1	$(z[-7,4,0],z[0,-5,3])$	$z[-14,20,25]$	12
$\Pi_{16}^{[0]}$	0	$(z[-9,12,-4],z[2,-10,5])$	$z[-14,20,25]$	24
$\Pi_{17}^{[0]}$	0	$(z[10,14,-11],z[8,-8,2])$	$z[-101,146,182]$	84
$\Pi_{18}^{[0]}$	0	$(z[14,28,-20],z[8,-8,2])$	$z[-510,737,919]$	368
$\Pi_{19}^{[0]}$	0	$(z[184,67,-94],z[8,-8,2])$	$z[-1656,2393,2984]$	1183
$\Pi_{20}^{[0]}$	0	$(z[-25,10,2],z[8,-8,2])$	$z[-227,328,409]$	149
$\Pi_{21}^{[0]}$	1	$(z[13,-9,1],z[6,6,-5])$	$z[-56,81,101]$	73
$\Pi_{22}^{[0]}$	1	$(z[-5,-8,6],z[-3,2,0])$	$z[-25,36,45]$	27
$\Pi_{23}^{[0]}$	1	$(z[43,14,-21],z[-3,11,-5])$	$z[-510,737,919]$	3072
$\Pi_{24}^{[0]}$	0	$(z[6,9,-7],z[4,-4,1])$	$z[-56,81,101]$	165
$\Pi_{25}^{[0]}$	0	$(z[-3,14,-7],z[4,-13,6])$	$z[-31,45,56]$	46
$\Pi_{26}^{[0]}$	0	$(z[-8,-3,4],z[-2,3,-1])$	$z[-25,36,45]$	20
$\Pi_{27}^{[0]}$	0	$(z[-1,-5,3],z[-4,6,-2])$	$z[-8,11,14]$	6

Table A.4 Continuation of Table A.3.

$\Pi_j^{[n]}$	v	**d**	ω	$\tilde{T}_j^{[n]}$
$\Pi_0^{[1]}$	1	$(z[-9,-2,4],z[-4,-1,2])$	$z[-2,3,3]$	1
$\Pi_1^{[1]}$	0	$(z[-30,-14,17],z[-25,-4,10])$	$z[-81,117,146]$	77
$\Pi_2^{[1]}$	1	$(z[-12,-4,6],z[-6,0,2])$	$z[-9,13,16]$	5
$\Pi_3^{[1]}$	1	$(z[-20,-5,9],z[-9,-2,4])$	$z[-3,4,6]$	1
$\Pi_4^{[1]}$	1	$(z[42,11,-19],z[-3,0,1])$	$z[-65,94,117]$	37
$\Pi_5^{[1]}$	0	$(z[22,4,-9],z[3,2,-2])$	$z[-7,10,13]$	1
$\Pi_6^{[1]}$	0	$(z[28,6,-12],z[3,2,-2])$	$z[-36,52,65]$	3
$\Pi_7^{[1]}$	1	$(z[-31,-8,14],z[-14,-3,6])$	$z[-16,23,29]$	1
$\Pi_8^{[1]}$	1	$(z[-32,-11,16],z[0,2,-1])$	$z[-81,117,146]$	2
$\Pi_0^{[2]}$	1	$(z[2,1,-1],z[1,0,0])$	$z[-1,2,2]$	1
$\Pi_1^{[2]}$	0	$(z[16,9,-10],z[7,-2,-1])$	$z[-20,29,36]$	6
$\Pi_2^{[2]}$	1	$(z[-4,-3,3],z[-3,4,-1])$	$z[-25,36,45]$	32
$\Pi_3^{[2]}$	0	$(z[-4,2,0],z[1,-4,2])$	$z[-3,4,5]$	2
$\Pi_4^{[2]}$	0	$(z[-8,-5,5],z[-5,1,1])$	$z[-25,36,45]$	2
$\Pi_5^{[2]}$	1	$(z[7,-2,-1],z[2,3,-2])$	$z[-11,16,20]$	4
$\Pi_0^{[3]}$	1	$(z[2,1,-1],z[1,0,0])$	$z[-1,2,2]$	1
$\Pi_1^{[3]}$	0	$(z[15,4,-7],z[8,1,-3])$	$z[-20,29,36]$	34
$\Pi_2^{[3]}$	0	$(z[46,10,-20],z[17,5,-8])$	$z[-36,52,65]$	2
$\Pi_3^{[3]}$	0	$(z[-198,-51,89],z[17,5,-8])$	$z[-263,380,474]$	22
$\Pi_4^{[3]}$	0	$(z[-50,-16,24],z[-5,1,1])$	$z[-81,117,146]$	8
$\Pi_5^{[3]}$	0	$(z[-12,-1,4],z[-1,-3,2])$	$z[-9,13,16]$	2
$\Pi_6^{[3]}$	1	$(z[-44,-17,23],z[-20,-1,7]$	$z[-81,117,146]$	64
$\Pi_7^{[3]}$	1	$(z[11,3,-5],z[1,0,0])$	$z[-4,6,7]$	2
$\Pi_8^{[3]}$	1	$(z[5,1,-2],z[1,0,0])$	$z[-2,3,4]$	2
$\Pi_9^{[3]}$	1	$(z[-14,-3,6],z[1,0,0])$	$z[-9,13,16]$	2

Table A.5 Incidence matrices of the $\pi/7$ model.

$$\mathbf{A}(0)=\begin{pmatrix}0&3&0&0&0&0&0&0&0&0&0&0\\0&0&0&0&0&0&0&0&0&0&0&0\\0&0&0&0&0&0&0&0&0&0&0&0\\0&2&1&0&0&0&0&0&0&0&0&0\\0&2&1&0&0&0&0&0&0&0&0&0\\2&0&0&2&1&0&3&0&4&10&9&0\\1&1&1&1&1&1&1&1&1&1&1&1\\0&0&0&0&0&0&0&0&0&0&0&0\\0&0&0&0&0&0&0&0&3&6&6&0\\0&0&0&1&0&0&2&1&1&7&6&4\\0&0&0&1&0&0&2&1&1&7&6&4\end{pmatrix}\quad \mathbf{A}(1)=\begin{pmatrix}0&0&0&0&0&0&0&0&0&0&0\\3&9&21&27&27&33&27&63&15&21&21\\0&6&17&22&22&27&22&54&12&17&17\\0&0&0&0&0&0&0&0&0&0&0\\0&0&0&0&0&0&0&0&0&0&0\\0&0&0&0&0&0&0&0&0&0&0\\0&0&0&0&0&0&0&0&0&0&0\\1&2&4&5&5&6&5&11&3&4&4\\0&0&0&0&0&0&0&0&0&0&0\\0&0&0&0&0&0&0&0&0&0&0\\0&0&0&0&0&0&0&0&0&0&0\end{pmatrix}$$

$$\mathbf{A}(2)=\begin{pmatrix}6&24&57&72&72&87&72&168&42&57&57\\0&0&0&0&0&0&0&0&0&0&0\\1&2&4&5&5&6&5&11&3&4&4\\1&5&12&15&15&18&15&35&9&12&12\\1&5&12&15&15&18&15&35&9&12&12\\1&3&7&9&9&11&9&21&5&7&7\\0&0&0&0&0&0&0&0&0&0&0\\0&0&0&0&0&0&0&0&0&0&0\\0&0&0&0&0&0&0&0&0&0&0\\0&0&0&0&0&0&0&0&0&0&0\\0&0&0&0&0&0&0&0&0&0&0\end{pmatrix}\quad \mathbf{A}(3)=\begin{pmatrix}0&0&0&0&0&0&0&0&0&0&0\\2&8&19&24&24&29&24&56&14&19&19\\0&0&0&0&0&0&0&0&0&0&0\\0&3&8&10&10&12&10&24&6&8&8\\1&3&7&9&9&11&9&21&5&7&7\\0&0&0&0&0&0&0&0&0&0&0\\1&2&4&5&5&6&5&11&3&4&4\\0&0&0&0&0&0&0&0&0&0&0\\0&0&0&0&0&0&0&0&0&0&0\\0&0&0&0&0&0&0&0&0&0&0\\0&0&0&0&0&0&0&0&0&0&0\\0&0&0&0&0&0&0&0&0&0&0\end{pmatrix}$$

$$\mathbf{A}(4)=\begin{pmatrix}4&16&38&48&48&58&48&112&28&38&38\\0&0&0&0&0&0&0&0&0&0&0\\0&0&0&0&0&0&0&0&0&0&0\\3&9&21&27&27&33&27&63&15&21&21\\0&5&13&16&16&19&16&38&10&13&13\\0&0&0&0&0&0&0&0&0&0&0\\0&0&0&0&0&0&0&0&0&0&0\\1&2&4&5&5&6&5&11&3&4&4\\0&0&0&0&0&0&0&0&0&0&0\\0&0&0&0&0&0&0&0&0&0&0\end{pmatrix}\quad \mathbf{A}(5)=\begin{pmatrix}1&2&4&5&5&6&5&11&3&4&4\\0&0&0&0&0&0&0&0&0&0&0\\0&0&0&0&0&0&0&0&0&0&0\\0&0&0&0&0&0&0&0&0&0&0\\0&0&0&0&0&0&0&0&0&0&0\\0&0&0&0&0&0&0&0&0&0&0\\0&0&0&0&0&0&0&0&0&0&0\\0&0&0&0&0&0&0&0&0&0&0\\1&1&1&1&1&1&1&1&1&1&1\\0&1&2&2&2&2&2&4&2&2&2\\0&0&1&2&2&3&2&6&0&1&1\\0&0&0&0&0&0&0&0&0&0&0\end{pmatrix}$$

Table A.6 Continuation of Table A.5.

$$\mathbf{A}(6) = \begin{pmatrix} 1 & 2 & 2 & 4 & 16 & 8 & 11 & 24 & 22 & 55 \\ 0 & 0 & 0 & 0 & 0 & 0 & 0 & 0 & 0 & 0 \\ 0 & 0 & 0 & 0 & 0 & 0 & 0 & 0 & 0 & 0 \\ 0 & 0 & 0 & 0 & 0 & 0 & 0 & 0 & 0 & 0 \\ 0 & 0 & 0 & 0 & 0 & 0 & 0 & 0 & 0 & 0 \\ 0 & 0 & 0 & 3 & 12 & 6 & 6 & 18 & 12 & 30 \\ 1 & 1 & 1 & 1 & 1 & 1 & 1 & 1 & 1 & 1 \\ 0 & 0 & 0 & 0 & 0 & 0 & 0 & 0 & 0 & 0 \\ 0 & 1 & 0 & 0 & 3 & 1 & 4 & 5 & 9 & 24 \\ 0 & 0 & 1 & 0 & 0 & 0 & 0 & 0 & 0 & 0 \end{pmatrix} \quad \mathbf{A}(7) = \begin{pmatrix} 4 & 16 & 38 & 48 & 48 & 58 & 48 & 112 & 28 & 38 & 38 \\ 0 & 0 & 0 & 0 & 0 & 0 & 0 & 0 & 0 & 0 & 0 \\ 0 & 0 & 0 & 0 & 0 & 0 & 0 & 0 & 0 & 0 & 0 \\ 3 & 9 & 21 & 27 & 27 & 33 & 27 & 63 & 15 & 21 & 21 \\ 0 & 5 & 13 & 16 & 16 & 19 & 16 & 38 & 10 & 13 & 13 \\ 0 & 0 & 0 & 0 & 0 & 0 & 0 & 0 & 0 & 0 & 0 \\ 0 & 0 & 0 & 0 & 0 & 0 & 0 & 0 & 0 & 0 & 0 \\ 1 & 2 & 4 & 5 & 5 & 6 & 5 & 11 & 3 & 4 & 4 \\ 0 & 0 & 0 & 0 & 0 & 0 & 0 & 0 & 0 & 0 & 0 \\ 0 & 0 & 0 & 0 & 0 & 0 & 0 & 0 & 0 & 0 & 0 \end{pmatrix}$$

$$\mathbf{A}(8) = \begin{pmatrix} 3 & 11 & 25 & 31 & 31 & 37 & 31 & 71 & 19 & 25 & 25 \\ 0 & 4 & 10 & 12 & 12 & 14 & 12 & 28 & 8 & 10 & 10 \\ 1 & 3 & 7 & 9 & 9 & 11 & 9 & 21 & 5 & 7 & 7 \\ 0 & 0 & 0 & 0 & 0 & 0 & 0 & 0 & 0 & 0 & 0 \\ 0 & 0 & 0 & 0 & 0 & 0 & 0 & 0 & 0 & 0 & 0 \\ 0 & 0 & 0 & 0 & 0 & 0 & 0 & 0 & 0 & 0 & 0 \\ 0 & 0 & 0 & 0 & 0 & 0 & 0 & 0 & 0 & 0 & 0 \\ 0 & 0 & 0 & 0 & 0 & 0 & 0 & 0 & 0 & 0 & 0 \\ 0 & 0 & 0 & 0 & 0 & 0 & 0 & 0 & 0 & 0 & 0 \\ 0 & 0 & 0 & 0 & 0 & 0 & 0 & 0 & 0 & 0 & 0 \end{pmatrix}$$

$$\mathbf{A}(9) = \begin{pmatrix} 8 & 23 & 44 & 23 & 113 & 53 & 98 & 173 & 203 & 518 \\ 8 & 23 & 44 & 23 & 113 & 53 & 98 & 173 & 203 & 518 \\ 2 & 7 & 16 & 5 & 29 & 13 & 28 & 45 & 59 & 152 \\ 2 & 7 & 16 & 5 & 29 & 13 & 28 & 45 & 59 & 152 \end{pmatrix}$$

$$\mathbf{A}(10) = \begin{pmatrix} 15 & 77 & 183 & 227 & 227 & 271 & 227 & 527 & 139 & 183 & 183 \\ 15 & 77 & 183 & 227 & 227 & 271 & 227 & 527 & 139 & 183 & 183 \\ 5 & 25 & 59 & 73 & 73 & 87 & 73 & 169 & 45 & 59 & 59 \\ 5 & 25 & 59 & 73 & 73 & 87 & 73 & 169 & 45 & 59 & 59 \end{pmatrix}$$

$$\mathbf{A}(11) = \begin{pmatrix} 21 & 69 & 150 & 93 & 453 & 213 & 357 & 693 & 741 & 1893 \\ 24 & 79 & 172 & 105 & 513 & 241 & 406 & 785 & 843 & 2154 \\ 10 & 33 & 72 & 43 & 211 & 99 & 168 & 323 & 349 & 892 \\ 10 & 33 & 72 & 43 & 211 & 99 & 168 & 323 & 349 & 892 \end{pmatrix}$$

$$\mathbf{A}(12) = \begin{pmatrix} 111 & 519 & 1227 & 1527 & 1527 & 1827 & 1527 & 3543 & 927 & 1227 & 1227 \\ 125 & 585 & 1383 & 1721 & 1721 & 2059 & 1721 & 3993 & 1045 & 1383 & 1383 \\ 51 & 239 & 565 & 703 & 703 & 841 & 703 & 1631 & 427 & 565 & 565 \\ 51 & 239 & 565 & 703 & 703 & 841 & 703 & 1631 & 427 & 565 & 565 \end{pmatrix}$$

$$\mathbf{A}(13) = \begin{pmatrix} 18 & 60 & 132 & 72 & 360 & 168 & 294 & 552 & 612 & 1566 \\ 16 & 53 & 116 & 67 & 331 & 155 & 266 & 507 & 553 & 1414 \\ 7 & 23 & 50 & 31 & 151 & 71 & 119 & 231 & 247 & 631 \\ 7 & 23 & 50 & 31 & 151 & 71 & 119 & 231 & 247 & 631 \end{pmatrix}$$

$$\mathbf{A}(14) = \begin{pmatrix} 84 & 396 & 936 & 1164 & 1164 & 1392 & 1164 & 2700 & 708 & 936 & 936 \\ 79 & 371 & 877 & 1091 & 1091 & 1305 & 1091 & 2531 & 663 & 877 & 877 \\ 37 & 173 & 409 & 509 & 509 & 609 & 509 & 1181 & 309 & 409 & 409 \\ 37 & 173 & 409 & 509 & 509 & 609 & 509 & 1181 & 309 & 409 & 409 \end{pmatrix}$$

A.5 Inadmissibility Tables for Models II and III

Model II, Sector A

(j_k,t_k)	$(7,1)$	$(7,2)$	$(10,4)$	$(7,0)$	$(22,0),(22,2)$	$(10,2)$	$(19,2)$	$(21,2)$
(j_{k+1},t_{k+1})	$(7,2)$	$(7,2),(10,4)$	$(7,1)$	$(22,2)$	$(7,0),(22,0)$	$(10,2)$	$(19,2)$	$(21,2)$

Model II, Sector B

(j_k,t_k)	(j_{k+1},t_{k+1})
$(0,0)$	$(6,3)$
$(3,0),(3,1)$	$(3,3)'$
$(3,3),(3,4)$	$(3,1)'$
$(3,0)',(3,2),(3,3)'$	$(3,1),(5,1),(6,1)$
$(3,1)',(3,2)',(3,4)'$	$(3,3),(5,5),(6,6)$
$(5,3),(5,4),(5,6)$	$(3,4)',(5,6),(8,1),(10,1)$
$(5,0),(5,1),(5,5),(5,6)'$	$(3,4),(5,6)',(6,7),(10,2)$
$(6,0),(6,1),(6,6),(6,7)$	$(0,0),(3,0),(5,0),(6,0)$
$(6,0)',(6,2),(6,3),(6,4)$	$(3,0)',(6,0)',(8,0)$
$(8,0)$	$(3,2)',(5,4),(6,4)$
$(8,1)$	$(3,2),(6,2)$
$(10,1),(10,2)$	$(5,3)$

Model III

(j_k,t_k)	(j_{k+1},t_{k+1})
$(0,0),(0,1)$	$(0,1),(1,7)',(1,10),(1,15),(1,17),$ $(2,7),(2,15),(2,17),(2,25),(2,32)$
$(0,0)',(0,1)'$	$(0,1)',(1,1),(1,3),(1,11),(1,8),$ $(2,18),(2,26),(2,1),(2,8),(2,16)$
$(0,0)''$	$(1,12),(2,27),(2,17)$
$(1,16),(1,17)$	$(1,2)'$
$(1,1),(1,2)$	$(1,16)$
$(1,t), t=2',3,4,9-12$	$(1,16),(1,9),(1,14),(2,24),(2,31),(2,14)$
$(1,t), t=0,7',8,9',14,15,16$	$(1,2),(1,4),(1,9)',(2,19),(2,2),(2,9)$
$(2,t), t=18,19,24-27$	$(0,0)$
$(2,t), t=0,1,2,31,32$	$(0,0)',(1,0),(2,0)$
$(2,t), t=0',7-9,14-17$	$(0,0)'',(1,0),(2,0)'$

References

Kouptsov K.L., Lowenstein J.H., and Vivaldi F. (2002) Quadratic rational rotations of the torus and dual lattice maps, *Nonlinearity* **15**, 1795–1842.

Lowenstein J.H., Kouptsov K. L., and Vivaldi F. (2004) Recursive tiling and geometry of piecewise rotations by $\pi/7$, *Nonlinearity* **17**, 1–25.

Appendix B
The Codometer

In this appendix we describe in concrete terms the software device, the *codometer*, which we use in this book to implement the symbolic dynamics appropriate to recursive tiling. Our intent here is to give the reader an intuitive feeling for the relatively simple and concrete machinery which we have found so useful in our computer "experiments".

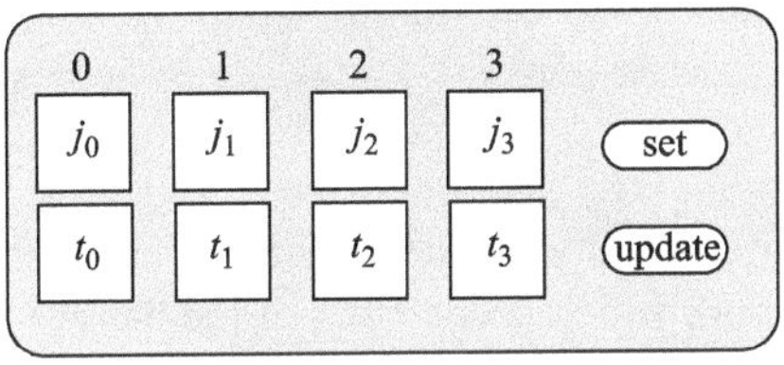

The general setup is shown in the figure. Our codometer consists of a linear array of *registers*, labeled $k = 0, 1, 2, \ldots K$, each assigned a pair of integers: a *cycle* $j_k \in J_k$ and a *phase* $t_k \in T_k(j_k) = \{0, 1, \ldots, \nu(j_k)\}$. We represent the *state* of the codometer, i.e. the set of current register values, by the expression

$$\left(\begin{pmatrix} j_0 \\ t_0 \end{pmatrix} \begin{pmatrix} j_1 \\ t_1 \end{pmatrix} \begin{pmatrix} j_2 \\ t_2 \end{pmatrix} \cdots \begin{pmatrix} j_K \\ t_K \end{pmatrix} \right).$$

The set J_k of possible cycles is assumed to be finite, but otherwise arbitrary. For each cycle j_k, the lower and upper limits for the corresponding phase t_k are stored in memory. Also stored in memory are the values, for $k > 0$, of the *predecessor functions* $p_k(j_k, t_k)$, which are essential both to set and to update the state of the codometer.

The codometer's inner workings only permit states for which the cycle and phase values neighboring registers satisfy the constraint

$$j_{k-1} = p_k(j_k, t_k).$$

Setting the state is initiated by entering j_K and t_K in the highest register. The codometer then supplies j_{K-1} using the predecessor constraint, while the user chooses t_{K-1} from the set $T_{K-1}(j_{K-1})$. This process continues until all the registers are set.

We now describe the updating process. If t_0 is less than its maximum value, we simply increment it to t_0+1. If, on the other hand, the first r phases are all maximal, but the $r+1$st is sub-maximal, then we increment t_r and reset the registers $r-1, r-2, \ldots, 0$, in that order, to obtain

$$\left(\binom{j'_0}{0}\binom{j'_1}{0}\cdots\binom{j'_{r-1}}{0}\binom{j_r}{t_{r+1}}\binom{j_{r+1}}{t_{r+1}}\cdots\binom{j_K}{t_K}\right).$$

where

$$j'_{r-1} = p_r(j_r, t_r+1), \qquad j'_{k-1} = p_k(j'_k, 0),\ k = r-1, r-2, \ldots, 1.$$

This updating rule uniquely specifies the stepwise incrementation of states, with the exception of the possibility that all t_k are maximal. The device may include such a rule in its program, or else restrict use to applications where the maximal state never arises (easily arranged in practice by choosing the number of registers, or the maximum value of the last register, large enough).

As a simple application, let us program a 4-register codometer as a "perpetual" Gregorian calendar, whose registers $k = 0,1,2,3$ are designed to keep track of days, months, years, and centuries, respectively. Following the usual convention, we will reset the registers for days and months to 1, and the registers for years and centuries to 0. This is a trivial modification easily incorporated into the codometer's program. For register 0, which keeps track of the days of the month, we have $j_0 = 28,29,30,31$, and t_0 runs from 1 to j_0, respectively. For register 1 (months of the year), $j_1 = 0$ (common year), 1 (leap year), with t_1 running from 1 to 12 in both cases. For register 2 (years of the century), we have $j_2 = 0,1$ corresponding to whether the century phase t_3 is divisible by 4 or not, with t_2 running from 0 to 99. Finally, register 3 simply counts the centuries: $j_3 = 0, t_3 \geq 0$. The upper limit on t_3 could be whatever limit fits our concept of a "perpetual" calendar. To complete the programming of the codometer, we need to specify the predecessor functions p_1, p_2, p_3 consistent with the dictates of Pope Gregory. First, we have

$$\{p_1(j_1,t_1) : t_1 = 1,\ldots,12\} = \{31, 28+j_1, 31, 30, 31, 30, 31, 31, 30, 31, 30, 31\}.$$

Then, $p_2(j_2,t_2) = 0$ if $j_2 = 0$ and $t_2 = 0$, or t_2 is not divisible by 4; otherwise, $p_2(j_2,t_2) = 1$. Finally, $p_3(j_3,t_3) = 0$ if t_3 is divisible by 4; otherwise, $p_3(j_3,t_3) = 1$.

As an example, consider the date December 31, 1999. To enter the register data, we first set $j_3 = 0$ (no choice) and t_3 to the number of the century, 19. Since 19 is not divisible by 4, this forces $j_2 = p_3(0,19) = 0$. To specify the year, we set $t_2 = 99$. Since the latter is not divisible by 4, the predecessor constraint forces $j_1 = p_2(0,99) = 0$, corresponding to a common year. Next we insert the month, $t_1 = 12$, which forces $j_0 = p_1(0,12) = 31$. Finally, we choose the day, $t_0 = 31$, and we are done. The resulting state is

$$\left(\begin{pmatrix}31\\31\end{pmatrix}\begin{pmatrix}0\\12\end{pmatrix}\begin{pmatrix}1\\99\end{pmatrix}\begin{pmatrix}0\\19\end{pmatrix}\right).$$

Updating by one day, to January 1, 2000, we follow the prescription described above (details left as an exercise for the reader) to obtain

$$\left(\begin{pmatrix}31\\1\end{pmatrix}\begin{pmatrix}1\\1\end{pmatrix}\begin{pmatrix}0\\0\end{pmatrix}\begin{pmatrix}0\\20\end{pmatrix}\right).$$

The codometers used in this book to study the aperiodic orbits of piecewise isometries, kicked oscillators, and nondissipative round-off maps have infinitely many registers, at least in theory (in computer simulations, of course, we deal only with a finite resolution, hence a finite number of levels of a scaling hierarchy, modeled by a codometer with finitely many registers). What makes this situation manageable is the fact that there are only a small finite set $\mathscr{I}$ of possible register *types*. In a codometer, the kth register is assigned a fixed index $i_k \in \mathscr{I}$, and the predecessor constraint does not depend on k, only on i_k. For the quadratic models discussed in this book, the number of distinct register types is typically 2 or 3: the zero register is distinguished from the rest, which, for a single scaling sequence, can be taken to be of the same type. For the cubic $\pi/7$ model, there are 15 distinct types, and each codometer has infinitely many registers and is assigned a fixed index string $\mathbf{i} = i_0 i_1 i_2 \ldots$. In the book, we represent its state as

$$\left(\begin{pmatrix}i_0\\j_0\\t_0\end{pmatrix}\begin{pmatrix}i_1\\j_1\\t_1\end{pmatrix}\begin{pmatrix}i_2\\j_2\\t_2\end{pmatrix}\begin{pmatrix}i_3\\j_3\\t_3\end{pmatrix}\cdots\right).$$

Index

Color Figure Index

Tiling of $D^{(\mathbf{A})}(0)$

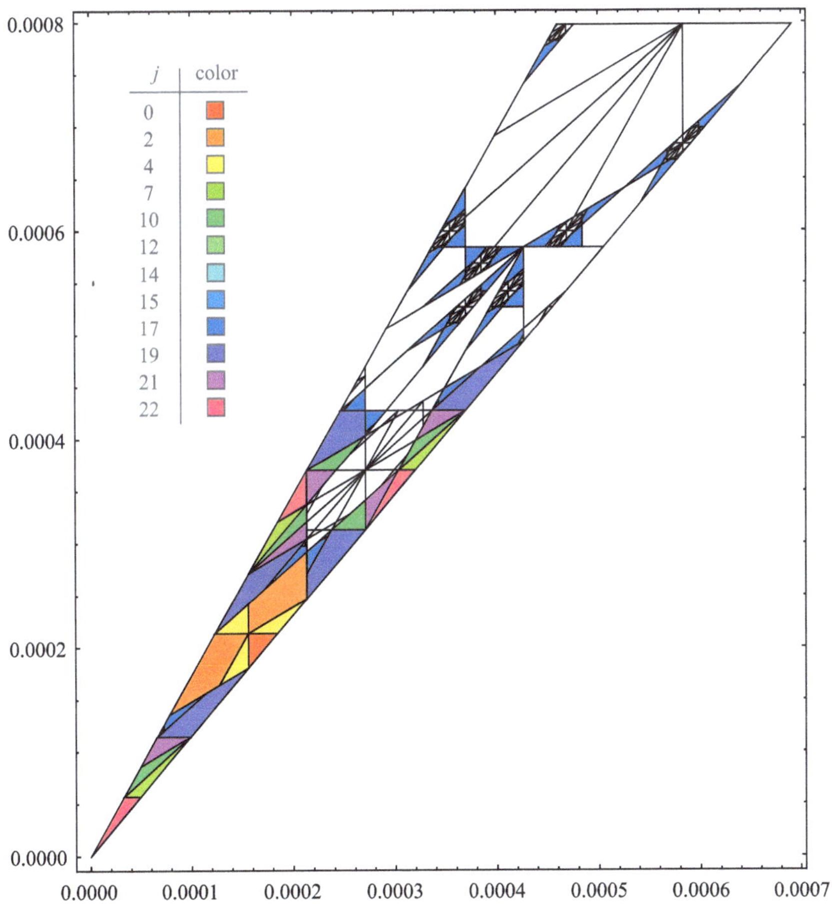

Color figure of page 45

Tiling of $D^{(\mathbf{B})}(0)$

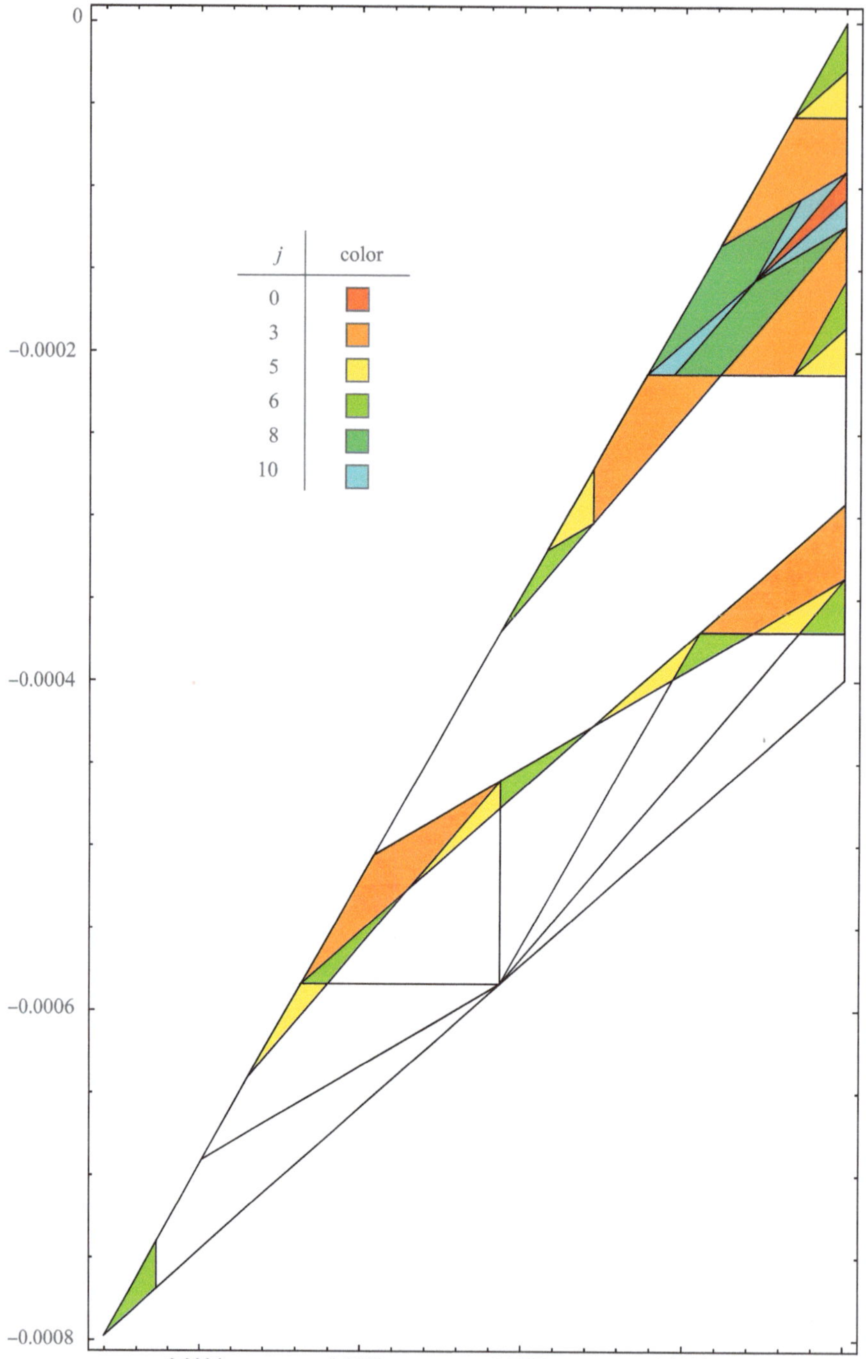

Color figure of page 45

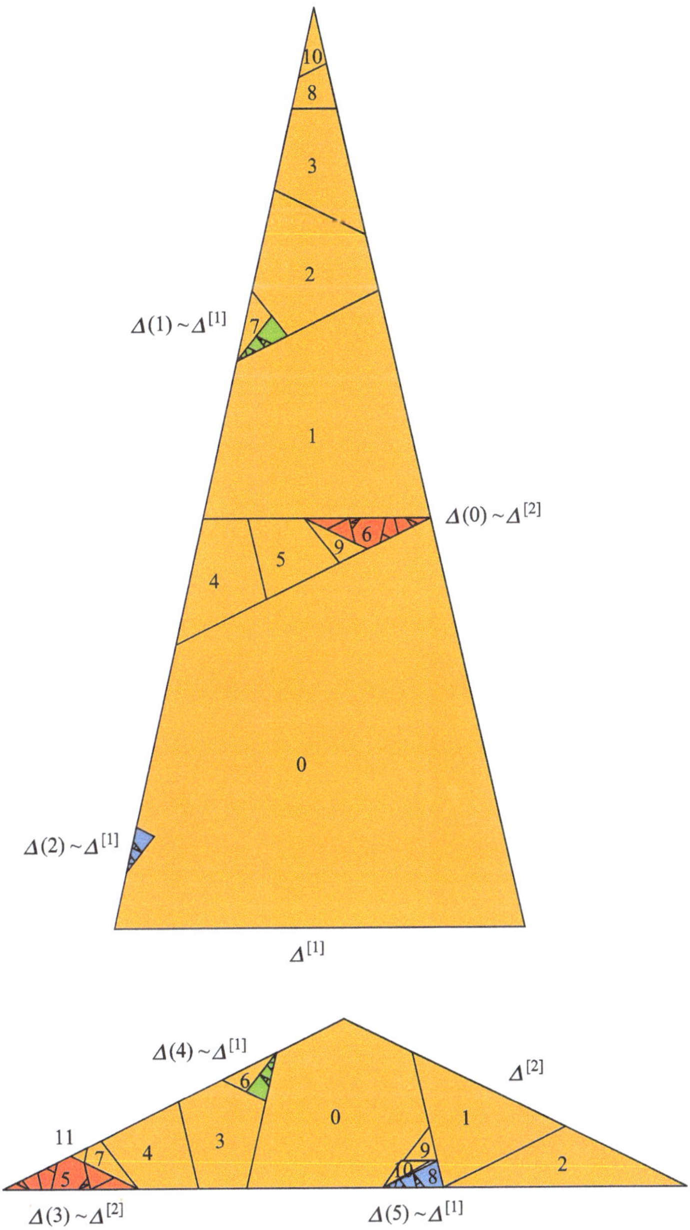

Fig. 2.18 Return map partitions of catalogue members $D^{[1]}$ and $D^{[2]}$ into their respective atoms. The Euclidean metric is used here to make evident the various similarity relations. Superposed are the dressed subdomains $\Delta(i)$, $i = 0, \ldots, 5$.

Fig. 2.19 Tiling of $D^{[1]}$ by the return orbits of $D_j(i)$, $i = 0, 1, 2$, $j = 0, \ldots, J(h(i))$. The Euclidean metric is used to make evident the similarity relations. Consistent with the previous figure, different values $i = 0, 1, 2$ are coded red, green, and blue, respectively.

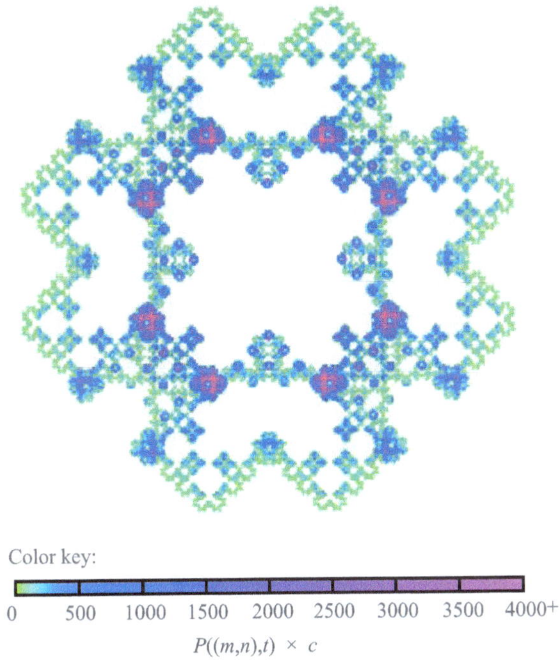

Fig. 6.7 Probability distribution $P((m,n),t)$, multiplied by the normalization factor $c = 5298000$, for web B, $-1000 \leq m,n < 1000, t = 72000000$.

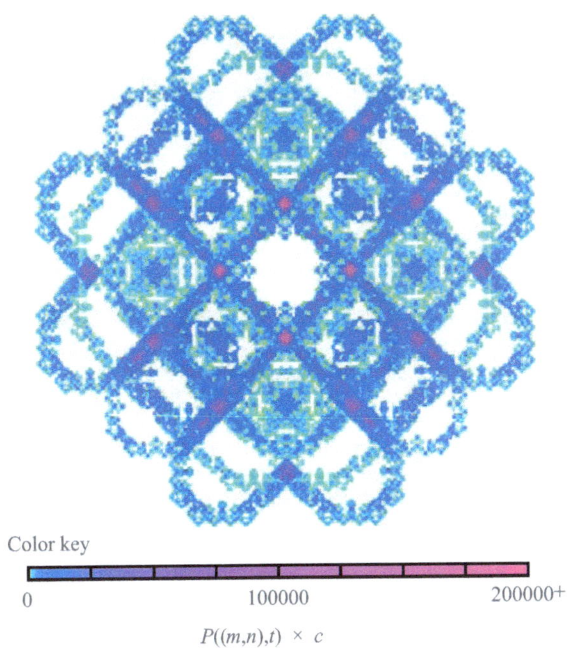

Fig. 6.8 Probability distribution $P((m,n),t)$, multiplied by the normalization factor $c = 161280$, for web A, $-550 \leq m,n < 550, t = 9000000$.

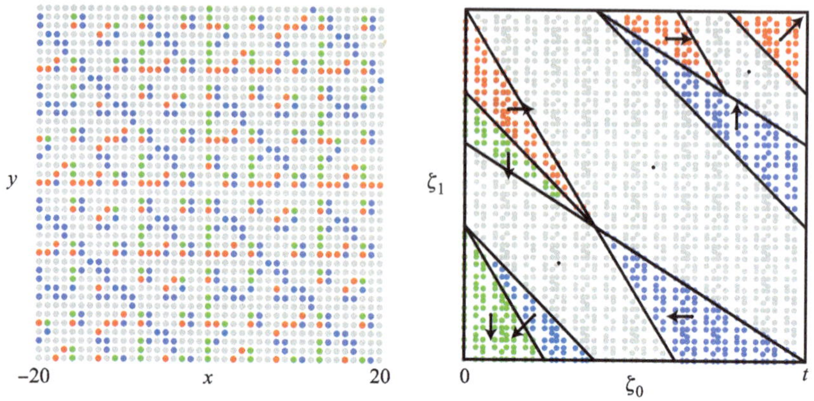

Fig. 7.5 Localization of the vector field for $p/q = 1/5$.

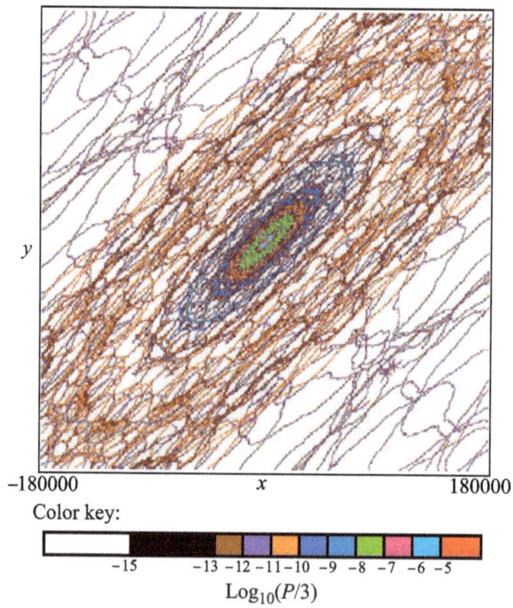

Fig. 7.20 Coarse-grained probability distribution for the round-off error for a time interval $t = 36000000$. There are 600×600 pixels, each covering 600^2 lattice sites. Colors are assigned to the indicated ranges of the total probability P per pixel.

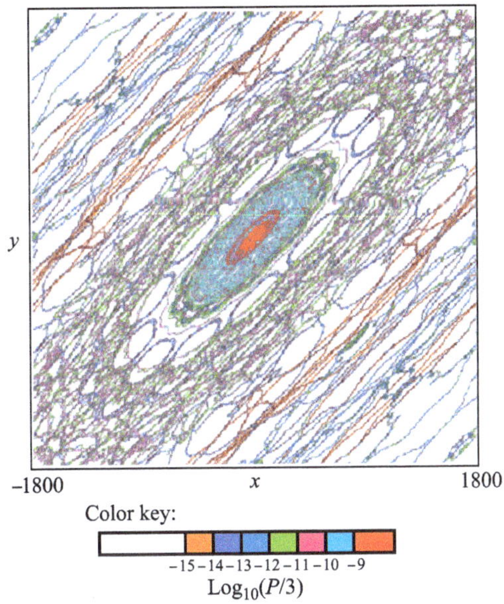

Fig. 7.21 Coarse-grained probability distribution for the round-off error for a time interval $t = 36000000$. There are 600×600 pixels, each covering 6^2 lattice sites. Colors are assigned to the indicated ranges of the total probability P per pixel.

FSC
www.fsc.org
MIX
Papier | Fördert gute Waldnutzung
FSC® C083411

Kolibri 360 GmbH
Ferdinand-Jühlke-Straße 7
99095 Erfurt, Deutschland
produktsicherheit@kolibri360.de